Conference Board of the Mathematical Sciences
REGIONAL CONFERENCE SERIES IN MATHEMATICS

supported by the
National Science Foundation

Number 30

BANACH SPACES OF ANALYTIC FUNCTIONS AND ABSOLUTELY SUMMING OPERATORS

by

ALEKSANDER PELCZYNSKI

Published for the
Conference Board of the Mathematical Sciences
by the
American Mathematical Society
Providence, Rhode Island

Expository Lectures
from the CBMS Regional Conference
held at Kent State University
July 11–16, 1976

AMS (MOS) subject classifications (1970).
Primary 46E15, 46J15; Secondary 30A98, 32E25.

Library of Congress Cataloging in Publication Data CIP

Pełczyński, Aleksander.
Banach spaces of analytic functions and absolutely summing operators.

(Regional conference series in mathematics; no. 30)
"Expository lectures from the CBMS regional conference held at Kent State University, July 11–16, 1976."
Includes bibliographical references.
1. Banach spaces. 2. Banach algebras. 3. Operator theory. 4. Holomorphic functions. I. Conference Board of the Mathematical Sciences. II. Title. III. Series.
QA1.R33 no. 30 [QA322.2] 510′.8s [515′.73] ISBN 0-8218-1680-2 77-9884

CONTENTS

Other Monographs in this Series

No. 1. Irving Kaplansky: *Algebraic and analytic aspects of operator algebras*
2. Gilbert Baumslag: *Lecture notes on nilpotent groups*
3. Lawrence Markus: *Lectures in differentiable dynamics*
4. H.S.M. Coxeter: *Twisted honeycombs*
5. George W. Whitehead: *Recent advances in homotopy theory*
6. Walter Rudin: *Lectures on the edge-of-the-wedge theorem*
7. Yozo Matsushima: *Holomorphic vector fields on compact Kähler manifolds*
8. Peter Hilton: *Lectures in homological algebra*
9. I. N. Herstein: *Notes from a ring theory conference*
10. Branko Grünbaum: *Arrangements and spreads*
11. Irving Glicksberg: *Recent results on function algebras*
12. Barbara L. Osofsky: *Homological dimensions of modules*
13. Michael Rabin: *Automata on infinite objects and Church's problem*
14. Sigurdur Helgason: *Analysis on Lie groups and homogeneous spaces*
15. R. G. Douglas: *Banach algebra techniques in the theory of Toeplitz operators*
16. Joseph L. Taylor: *Measure algebras*
17. Louis Nirenberg: *Lectures on linear partial differential equations*
18. Avner Friedman: *Differential games*
19. Béla Sz.-Nagy: *Unitary dilations of Hilbert space operators and related topics*
20. Hyman Bass: *Introduction to some methods of algebraic K-theory*
21. Wilhelm Stoll: *Holomorphic functions of finite order in several complex variables*
22. O. T. O'Meara: *Lectures on linear groups*
23. Mary Ellen Rudin: *Lectures on set theoretic topology*
24. Melvin Hochster: *Topics in the homological theory of modules over commutative rings*
25. Karl. W. Gruenberg: *Relation modules of finite groups*
26. Irving Reiner: *Class groups and Picard groups of group rings and orders*
27. H. Blaine Lawson, Jr.: *The quantitative theory of foliations*
28. T. A. Chapman: *Lectures on Hilbert cube manifolds*
29. Alan Weinstein: *Lectures on symplectic manifolds*
30. Aleksander Pelczynski: *Banach spaces of analytic functions and absolutely summing operators*
31. Ronald R. Coifman and Guido Weiss: *Transference methods in analysis*
32. Wolfgang M. Schmidt: *Small fractional parts of polynomials*

Preface

This is a revised version of my ten-lecture marathon at the American Mathematical Society Regional Conference "On Banach spaces of analytic functions" held at Kent State University, July 11–16, 1976. It is based on the course which I taught at the Ohio State University during the fall and winter quarters of the academic year 1975–1976.

In the past, the "classical" spaces of analytic functions, the disc algebra A and the Hardy spaces H^p, have been studied mainly from the point of view of the theory of Banach algebras and harmonic analysis. However, several results obtained more recently can be naturally formulated in the language of general Banach spaces and operator theory, although they are still closely related to some questions in hard analysis and often use analytic tools. An example is the Fefferman-Stein duality between BMO and H^1 (cf. **[F-S]**). Another one is Henkin's theory of analytic measures for spaces of analytic functions of several complex variables in strictly pseudoconvex domains (cf. **[He3]**). Also the techniques of p-absolutely summing operators and related ideas which have been recently the object of intense study in the theory of Banach spaces seem to offer a new approach to spaces of analytic functions. To explain this, note that the natural injection $A \longrightarrow H^p$ is an example of a p-absolutely summing operator, and the fact that this operator is p-integral for $p > 1$ is equivalent to the M. Riesz theorem on the boundedness in L^p-norm of the orthogonal projection from L^p onto H^p. More generally considering the completion in $L^p(\mu)$-norm of the uniformly continuous holomorphic functions in a fixed bounded domain of holomorphy, where μ is a measure on the boundary of the domain, is nothing more than studying certain properties of a certain p-absolutely summing operator: "natural injection". A more sophisticated example is a map $f \longrightarrow (\hat{f}(2^n))_{n=1}^{\infty}$ from A into l^2. An inequality discovered by Paley around 1932 says that

$$\left(\sum_{n=1}^{\infty} |\hat{f}(2^n)|^2\right)^{1/2} \leqslant 2 \int_{\partial D} |f|\, dm \quad \text{for } f \in A.$$

It means exactly that the map in question is absolutely summing. Paley's inequality depends heavily on the analyticity of f. There is no counterpart for continuous functions on the circle and this reflects the fact that every absolutely summing operator from a $C(S)$-space into a Hilbert space is nuclear, and therefore compact. This produces a linear topological invariant which enables us to show that the disc algebra (and more generally a uniform algebra with a nontrivial Gleason part) is not isomorphic, as a Banach space, to any quotient of a $C(S)$-space (cf. §§4 and 5). On the other hand, Paley's inequality can be applied to prove the basic fact in the theory of absolutely summing operators, discovered by Grothendieck, that every operator from an $L^1(\mu)$-space to a Hilbert space is absolutely summing (cf. §3 for details).

Some words about the organization of the paper:

§§0 and 1 have preliminary character. §§2 and 3 are devoted to studying properties of p-absolutely summing operators from the disc algebra. The main result proved there says that if $p > 1$ then every p-absolutely summing operator from the disc algebra is p-integral. §§4 and 5 should convince the reader that the disc algebra is a pathological space from the point of view of Banach space theory, or at least that it is very different from $C(S)$-spaces, and that the same is true for a large class of uniform algebras. On the other hand in §§7 and 8 we show that the disc algebra shares various properties of $C(S)$-spaces like the Dunford-Pettis property, characterizations of weakly compact operators, weak sequential completeness of the dual, etc. The basic analytic tool which leads to these results is the Havin lemma which is presented in §6. Roughly speaking, the Havin lemma is a quantitative analytic refinement of Fatou's theorem that closed sets in the unit circle, of Lebesgue measure zero, are the peak sets for the disc algebra. In §9 we apply the main result of §2 to study asymptotic behaviour of norms of projections from the disc algebra onto its finite dimensional subspaces as the dimensions of these subspaces tend to infinity. We prove that the dual of the disc algebra has Rademacher cotype p for some $p < \infty$. The crucial question, whether this dual has cotype 2, still remains open.

§10 is loosely related to other sections. It is a survey (without proofs) of results concerning bases and various approximation properties in the classical spaces of analytic functions.

In the last §11 we consider the problem of isomorphic classification of spaces of analytic functions of different numbers of variables. Most of the section is devoted to an elementary presentation of Henkin's generalization of the M. and F. Riesz theorem for the ball algebras.

The material presented here mostly concerns the disc algebra and its generalization to several complex variables. The Banach space properties of other H^p spaces are almost untouched. The reader interested in this subject is referred to the excellent survey by Coifman and Weiss **[C-W]** and to the paper **[Kw-P]**.

Finally I would like to express my gratitude to everybody who helped me to complete this work:

to Joseph Diestel who got the idea of organizing the Regional Conference in Kent and who was able to transform his dream into reality,

to Peter Ørno whose permanent interest in the subject and sharp criticism constantly stimulated my work; I am particularly indebted to him for permission to include several of his unpublished results,

to W. J. Davis, G. A. Edgar, T. Figiel, W. B. Johnson, G. Pisier and P. Wojtaszczyk who are responsible for many improvements in the text.

My sincere thanks go to Pam Walsh for her typing job and patience in preparing the manuscript.

I would like to also acknowledge the Ohio State University and the National Science Foundation for their financial support.

ALEKSANDER PELCZYNSKI
POLISH ACADEMY OF SCIENCES
OHIO STATE UNIVERSITY

0. Preliminaries

Our terminology and notation for Banach spaces and linear operators is standard; it is similar to that employed by Dunford and Schwartz **[D-SI]**, and Lindenstrauss and Tzafriri **[L-T2]**.

0.I. *$L^p(\mu)$-spaces and $C(S)$-spaces.* Let μ be a nonnegative (not necessarily sigma finite) measure defined on a sigma field of subsets of a set Ω. Let $0 < p \leqslant \infty$. By $L^p(\mu)$ (resp. $L^p_R(\mu)$) we denote the space of μ-equivalence classes of complex-valued (resp. real-valued) functions f such that $\|f\|_p < \infty$, where

$$\|f\|_p = \begin{cases} \int_\Omega |f|^p \, d\mu & \text{for } 0 < p < 1, \\ \left(\int_\Omega |f|^p \, d\mu \right)^{1/p} & \text{for } 1 \leqslant p < \infty, \\ \operatorname*{ess\,sup}_{s \in S} |f(s)| & \text{for } p = \infty. \end{cases}$$

For $1 \leqslant p \leqslant \infty$, $L^p(\mu)$ under the norm $\|\cdot\|_p$ is a Banach space.

By $C(S)$ (resp. $C_R(S)$) we denote the Banach space of all the continuous complex-valued (resp. real-valued) functions on a compact Hausdorff space S with the norm $\|f\| = \sup_{s \in S} |f(s)|$. Given an $f \in C(S)$ we define the functions $|f|$ and $\bar{f}$ by $|f|(s) = |f(s)|$, $\bar{f}(s) = \overline{f(s)}$ for $s \in S$. The constant functions are identified with the scalars.

We identify the dual space $[C(S)]^*$ via the Riesz representation theorem with the space of all complex Borel measures on S with the norm $\|\mu\| =$ the total variation of μ. We put $[C(S)]^*_+ = \{\mu \in [C(S)]^* : \mu \text{ nonnegative}\}$. A $\mu \in [C(S)]^*_+$ with $\mu(S) = \|\mu\| = 1$ is called a *probability measure.* Given a $\mu \in [C(S)]^*_+$ and a $\nu \in [C(S)]^*$ we write $\nu \ll \mu$ if ν is absolutely continuous with respect to μ, and $\nu \perp \mu$ if ν is singular with respect to μ. Let $\mu \in [C(S)]^*_+$. Then there is a natural isometric isomorphism from $L^1(\mu)$ onto the subspace $\{\nu \in [C(S)]^* : \nu \ll \mu\}$ which assigns to each $g \in L^1(\mu)$ the measure ν defined by

$$\nu(A) = \int_A g \, d\mu \quad \text{for every Borel set } A \subset S. \tag{0.1}$$

Given $\nu \ll \mu$, the unique g satisfying (0.1) is denoted by $d\nu/d\mu$ and it is called the *Radon-Nikodym derivative of* ν with respect to μ. For a $\nu \in [C(S)]^*$ we denote by $|\nu|$ the unique element of $[C(S)]^*_+$ such that $\nu \ll |\nu|$ and $|d\nu(s)/d|\nu|| = 1$ $|\nu|$-almost everywhere on S. Finally given a $\nu \in [C(S)]^*$ and a $g \in L^1(|\nu|)$ we denote by $g \cdot \nu$ the unique measure in $[C(S)]^*$ whose Radon-Nikodym derivative with respect to $|\nu|$ is $g \cdot (d\nu/d|\nu|)^{-1}$.

0.II. *Classical spaces of analytic functions and the Hilbert transform.* General references to this part are the books of Duren **[Du]**, Hoffman **[H]**, and the treatise of Zygmund **[Z]**.

In the sequel

C–denotes the complex plane,

$D = \{z \in \mathbf{C} : |z| \leqslant 1\}$–the closed unit disc,

$\partial D = \{z \in \mathbf{C} : |z| = 1\}$–the unit circle = the boundary of D,

m–denotes (except §11) the normalized Lebesgue measure on ∂D,

$L^p = L^p(m)$ and $L^p_R = L^p_R(m)$ $(0 < p \leqslant \infty)$,

A–stands for the *Disc Algebra* which is defined to be the smallest closed (in the topology of uniform convergence) subspace of $C(\partial D)$ which contains all the polynomials in z, $P(z) = \Sigma_{k=0}^n c_k z^k$ (c_k–arbitrary complex numbers, $k = 0, 1, \ldots, n$; $n = 0, 1, \ldots$).

For $0 \leqslant p < \infty$ we define the *Hardy space* H^p to be the subspace of L^p consisting of all the functions f which are limits of a sequence of polynomials (P_s) in the norm $\|\cdot\|_p$, i.e., $\lim_s \int_{\partial D} |P_s - f|^p \, dm = 0$.

For $p = \infty$ we put

$$H^\infty = \left\{ f \in L^\infty : \int_{\partial D} f(z) z^n \, m(dz) = 0 \quad \text{for } n = 1, 2, \ldots \right\}.$$

Next we put

$$A_0 = zA = \{f \in A : f = zg \text{ for some } g \in A\},$$

$$H^p_0 = zH^p = \{f \in H^p : f = zg \text{ for some } g \in H^p\}.$$

(Here z denotes the identity function on ∂D.)

Obviously A and H^p for $1 \leqslant p \leqslant \infty$ are complex Banach spaces; if $0 < p < 1$, then L^p and H^p are complete complex linear metric spaces. Clearly A can be identified with a closed linear subspace of H^∞ via the map which assigns to each $f \in A$ its m-equivalence class.

Let g be an analytic function in the open unit disc $D \backslash \partial D$ such that for m-almost all $z \in \partial D$ there exists the radial limit, $\lim_{r \uparrow 1} g(rz)$. Then the measurable function $z \longrightarrow \lim_{r \uparrow 1} g(rz)$ (defined m-almost everywhere on ∂D) is called the *boundary value function* of g.

Recall (cf. **[Du**, Chapters 2 and 3]) that every $f \in H^p$ $(0 < p \leqslant \infty)$ is a boundary value function of a unique analytic function in $D \backslash \partial D$; which is called the *analytic extension* of f; we denote this function, unless otherwise stated, also by f. It satisfies the inequality

$$\sup_{0 < r < 1} M_p(r, f) < \infty \tag{0.2}$$

where $M_p(r, f) = \int_{\partial D} |f(rz)|^p \, m(dz)$ for $0 < p < \infty$ and $M_\infty(r, f) = \operatorname{ess\,sup}_{z \in \partial D} |f(rz)|$.

Conversely, if f is an analytic function in $D \backslash \partial D$ which satisfies (0.2) for some p with $0 < p \leqslant \infty$, then there exists a boundary value function of f and it belongs to H^p.

An $f \in A$ iff f extends to a continuous function on the disc D which is analytic in $D \backslash \partial D$.

For $f \in L^1$ we put

$$\hat{f}(n) = \int_{\partial D} f(z) z^{-n}\, m(dz) \qquad (n = 0, \pm 1, \ldots).$$

If $f \in H^1$ and if we identify f with its analytic extension, then

(0.3) $$\hat{f}(n) = \frac{f^{(n)}(0)}{n!} \quad \text{for } n \geqslant 0.$$

Clearly if $f \in H^1$, then $f(n) = 0$ for $n < 0$. The converse implication is also true; it follows from

THEOREM 0.1 (THE FEJER THEOREM). *If $f \in L^p$ $(1 \leqslant p < \infty)$, then*

$$\lim_n \int_{\partial D} \left| f(z) - \frac{1}{n} \sum_{k=-n+1}^{n} (n - |k|) \hat{f}(k) z^k \right|^p m(dz) = 0.$$

For the proof cf. [**H**, p. 23]. □

We introduce now the operator H called the *Hilbert transform* which plays the crucial role in the study of "classical" spaces of analytic functions.

For $u \in L^1_R$ we define the Hilbert transform of u to be the unique function $v = H(u)$ such that $u + iv$ is a boundary value function of an analytic function f in $D \backslash \partial D$ such that $f(0) = \int_{\partial D} u\, dm$ and

$$\operatorname{Re} f(re^{i\vartheta}) = (2\pi)^{-1} \int_0^{2\pi} u(e^{i\vartheta}) \frac{1 - r^2}{1 - 2r\cos\vartheta + r^2}\, d\vartheta \quad \text{for } re^{i\vartheta} \in D \backslash \partial D.$$

For $g \in L^1$ the Hilbert transform $H(g)$ is defined by

$$H(g) = H(\operatorname{Re} g) + iH(\operatorname{Im} g)$$

where $\operatorname{Re} g = \frac{1}{2}(g + \bar{g})$. $\operatorname{Im} g = (g - \bar{g})/2i$.

The next result summarizes the most important properties of the Hilbert transform (for the proof cf. [**Du**, Chapter 4]).

THEOREM 0.2. (i) *If $\infty > p > 1$, then $H(f) \in L^p$ whenever $f \in L^p$; moreover there exists an absolute constant a independent of p such that*

$$\|H(f)\|_p \leqslant a \frac{p^2}{p-1} \|f\|_p \quad \textit{for every } f \in L^p$$

and

$$\|H(f)\|_p > a^{-1} \frac{p^2}{p-1} \|f\|_p \quad \textit{for some } f = f_p \in L^p.$$

(ii) (*the Kolmogorov theorem*) *If $f \in L^1$, then $H(f) \in L^p$ for every $p < 1$. Hence there exists a constant C_p such that*

$$\|H(f)\|_p \leqslant C_p \|f\|_1^p \quad \textit{for every } f \in L^1.$$

(iii) *The operator $f \longrightarrow H(f)$ is of weak type* (1-1), *i.e., there is an absolute constant C_1 such that for every $K > 0$*

0.III. *Absolutely summing operators and their relatives.* General references to this part are the papers by Grothendieck **[Gr1]**, **[Gr3]**, Pietsch **[Pi]**, Persson and Pietsch **[P-P]** and Seminaire Maurey-Schwartz 1972–1973, 1973–1974, 1974–1975 **[M-S]**.

$B(X, Y)$ stands for the space of all the bounded linear operators from a Banach space X into a Banach space Y.

Let $1 \leqslant p < \infty$. Let S be a compact Hausdorff space and let $\mu \in [C(S)]^*_+$. By $i_{\mu,p}$ we denote the natural injection of $C(S)$ into $L^p(\mu)$, i.e., the operator which assigns to each $f \in C(S)$ its μ-equivalence class regarded as an element of $L^p(\mu)$. If E is a closed linear subspace of $C(S)$, then (unless otherwise stated) $E_{\mu,p}$ denotes the subspace of $L^p(\mu)$ being the closure in $L^p(\mu)$-norm of $i_{\mu,p}(E)$ and $i^E_{\mu,p}$ denotes the restriction of $i_{\mu,p}$ to E regarded as an operator to $E_{\mu,p}$.

DEFINITION 0.1. Let $1 \leqslant p < \infty$. A bounded linear operator $T: X \longrightarrow Y$ is *p-integral* (resp. *strictly p-integral*) if there are a compact Hausdorff space S, a $\mu \in [C(S)]^*_+$ and bounded linear operators $U: X \longrightarrow C(S)$ and $V: L^p(\mu) \longrightarrow Y^{**}$ (resp. $V: L^p(\mu) \longrightarrow Y$) such that

$$\kappa_Y \cdot T = Vi_{\mu,p}U \quad (\text{resp. } T = Vi_{\mu,p}U) \tag{0.4}$$

where $\kappa_y: Y \longrightarrow Y^{**}$ denotes the canonical embedding of Y into its second dual Y^{**}.

A triple $(U, V, i_{\mu,p})$ satisfying (0.4) is called a *p-integral factorization* of T. The *p-integral norm* of T is the quantity

$$i_p(T) = \inf \|U\| \, \|V\| \, \|\mu\|^{1/p}$$

where the infimum is extended over all p-integral representations of T.

DEFINITION 0.2. Let $1 \leqslant p < \infty$. A bounded linear operator $T: X \longrightarrow Y$ is *p-absolutely summing* if there are a compact Hausdorff space S, a $\mu \in [C(S)]^*_+$, a closed linear subspace E of $C(S)$, and bounded linear operators $U: X \longrightarrow E$ and $V: E_{\mu,p} \longrightarrow Y$ such that

$$Vi^E_{\mu,p}U = T, \tag{0.5}$$

A triple $(U, V, i^E_{\mu,p})$ satisfying (0.5) is called a *p-absolutely summing factorization* of T. The *p-absolutely summing norm* of T is the quantity

$$\pi_p(T) = \inf \|U\| \, \|V\| \, \|\mu\|^{1/p}$$

where the infimum is extended over all p-absolutely summing factorizations of T. We shall use the term "absolutely summing" instead of "1-absolutely summing".

The following result is a slightly improved version of the so-called Grothendieck-Pietsch theorem (cf. **[Pi]**, **[P-P]**, **[Mt-P]**, **[P8]**).

THEOREM 0.4. *Let $1 \leqslant p < \infty$ and let $T \in B(X, Y)$.*

(i) *If T is p-integral and $j: X \longrightarrow C(S_0)$ is a fixed isometric isomorphic embedding of X into $C(S_0)$, S_0 arbitrary compact Hausdorff space, then there exists a $\mu \in [C(S_0)]^*_+$ and a linear operator $V: L^p(\mu) \longrightarrow Y^{**}$ such that*

$$i_p(T) = \|\mu\|^{1/p}, \quad \|V\| = 1, \quad \kappa_y T = Vi_{\mu,p}j.$$

(ii) *If T is p-absolutely summing and $j: X \longrightarrow C(S_0)$ is an isometric embedding of X into $C(S_0)$, S_0 arbitrary compact Hausdorff space, then there exists a $\mu \in [C(S_0)]^*_+$ and a linear operator $V: E_{\mu,p} \longrightarrow Y$ where $E = j(X)$ such that*

$$(0.6) \qquad \pi_p(T) = \|\mu\|^{1/p}, \quad \|V\| = 1, \quad T = Vi^E_{\mu,p}j.$$

(iii) *If there is a constant $C < \infty$ such that*

$$(0.7) \qquad \sum_{j=1}^{n} \|Tx_j\|^p \leqslant C^p \sup_{\|x^*\| \leqslant 1} \sum_{i=1}^{n} |x^*(x_j)|^p$$

for arbitrary $x_1, \ldots, x_n$ in X ($n = 1, 2, \ldots$),

then T is p-absolutely summing with $\pi_p(T) = \inf\{C: C$ satisfies (0.7)$\}$. Conversely, if T is p-absolutely summing, then T satisfies (0.7) with $C = \pi_p(T)$. □

Clearly every p-integral operator, say T, is p-absolutely summing and $\pi_p(T) \leqslant i_p(T)$. Using the fact that every closed linear subspace of a Hilbert space is a range of the norm one projection from the whole space, it can be easily seen that every 2-absolutely summing operator is (strictly) 2-integral. By Theorem 0.5(ii), every p-absolutely summing operator from a $C(S)$-space is (strictly) p-integral. Next observe that if Y is a dual Banach space (in particular Y is reflexive) or more generally if Y is complemented in its second dual, then every p-integral operator is strictly p-integral.

A discrete analogue of strictly p-integral operators are the p-nuclear operators which we define next.

DEFINITION 0.3. Let $1 \leqslant p < \infty$. A bounded linear operator $T: X \longrightarrow Y$ is *p-nuclear* if there are bounded linear operators $U: X \longrightarrow l^\infty$, $V: l^p \longrightarrow Y$, and a sequence of scalars (λ_j) with $\Sigma|\lambda_j|^p < \infty$ such that

$$T = V\Lambda_p U$$

where $\Lambda_p: l^\infty \longrightarrow l^p$ is defined by $\Lambda_p((\xi_i)) = (\lambda_i\xi_i)$ for $(\xi_i) \in l^\infty$. The triple (U, V, Λ_p) is called a *p-nuclear factorization* of T. The *p-nuclear norm* of T is the quantity

$$n_p(T) = \inf \|U\|\,\|V\|\,\|\Lambda_p\|$$

where the infimum is extended over all p-nuclear factorizations of T.

It is easy to see that every p-nuclear operator is compact.

For $p = 1$ we write $n(T)$ instead of $n_1(T)$ and we shall use the term "nuclear" instead of "1-nuclear". The set of nuclear operators from X into Y is denoted by $N(X, Y)$. It is a Banach space under the norm $n(\cdot)$.

We shall need the following result due to Grothendieck **[Gr3]**, cf. also **[D-U]**, and **[Per]** for the reflexive case.

THEOREM 0.5. *If Y is a Banach space with a Radon-Nikodym property, then every absolutely summing operator from a $C(S)$-space into Y is nuclear and therefore compact.*

$$m\{z \in \partial D\colon |(\mathcal{H}f(z))| \geq K\} \leq C_1 K^{-1} \int_{\partial D} |f|\, dm \quad \text{for } f \in L^1.$$

(iv) *If f satisfies the Lipschitz condition, then $\mathcal{H}(f)$ is continuous; if f is a C^∞-smooth function so is $\mathcal{H}(f)$.*

(v) *If $\mathcal{H}(f) \in L^1$, then*

$$\widehat{\mathcal{H}(f)}(n) = \begin{cases} -i\hat{f}(n) & \text{for } n > 0, \\ \hat{f}(0) & \text{for } n = 0, \\ i\hat{f}(n) & \text{for } n < 0. \end{cases} \qquad \square$$

The close relatives of the Hilbert transform are: the Riesz projection $\mathcal{R}$ and the Boas isomorphism $\mathcal{B}$. They are defined via the Hilbert transform by

$$\mathcal{R}(f) = 2^{-1}(f + i\mathcal{H}(f) + (1 - i)\hat{f}(0)) \qquad (f \in L^1),$$

$$2\mathcal{B}(f)(z) = [f + i\mathcal{H}(f)](z^2) + z^{-1}[f - i\mathcal{H}(f)](z^{-2}) + (1 - i)\hat{f}(0)(1 - 2^{-1}) \qquad (f \in L^1, z \in \partial D).$$

Theorem 0.2 yields (for the proof of (jjjj) cf. Boas **[Bo]** and **[K-P]**.

THEOREM 0.3. (j) *Let $1 < p < \infty$. Then $\mathcal{R}$ is a bounded projection from L^p onto its subspace H^p such that* $\ker \mathcal{R} = \{f \in L^p : \bar{f} \in H_0^p\}$. *Moreover, if $\|\mathcal{R}\|_p$ denotes the norm of this projection regarded as an operator on L^p, then*

$$b^{-1}\frac{p^2}{p-1} \leq \|\mathcal{R}\|_p \leq b\frac{p^2}{p-1}$$

where b is an absolute constant independent of p.

(jj) (*the Kolmogorov theorem*) *If $f \in L^1$, then $\mathcal{R}f \in H^p$ for every p with $0 < p < 1$. Hence there is a constant C_p such that*

$$\|\mathcal{R}f\|_p \leq C_p \|f\|_1^p \quad \text{for } f \in L^1 \qquad (0 < p < 1).$$

(jjj) *The operator $f \longrightarrow \mathcal{R}(f)$ is of weak type* (1-1).

(jjjj) *Let $1 < p < \infty$. Then $\mathcal{B}$ maps isomorphically L^p onto H^p; we have*

$$\mathcal{B}(f) = \hat{f}(0) + \sum_{n=1}^{\infty} [\hat{f}(n)z^{2n} + \hat{f}(-n)z^{2n-1}] \qquad (f \in L^p)$$

(*the series on the right-hand side converges in L^p*). *Moreover*

$$4^{-1}\|f\|_p \leq \|\mathcal{B}(f)\|_p \leq c\frac{p^2}{p-1}\|f\|_p \qquad (f \in L^p)$$

where c is an absolute constant independent of p.

Finally observe that sometimes it is more convenient to deal with the models of L^p and H^p which consist of 2π-periodic functions on the real line; in these models the Hilbert transform is given by the formula

$$\mathcal{H}(f)(t) = -\frac{1}{\pi}\lim_{\epsilon \to 0}\int_\epsilon^\pi \frac{[g(t+s) - g(t-s)]}{2tg(s/2)}\, ds$$

for m-almost all t.

Recall that a Banach space Y has the *Radon-Nikodym property* if for every finite measure μ on a sigma field of subsets of a set Ω every $V \in B(L^1(\mu), Y)$ admits a representation

$$V(f) = \int_\Omega f(\omega)y(\omega)\mu(d\omega), \qquad f \in L^1(\mu), \tag{0.8}$$

where $y(\cdot): \Omega \longrightarrow Y$ is a μ-Bochner integrable μ-essentially bounded function.

While the proof of Theorem 0.5 in full generality is slightly complicated, the fact that the operators in question are compact is easy. Indeed, if $T: C(S) \longrightarrow Y$ is absolutely summing then, by Theorem 0.4(ii), it admits a factorization

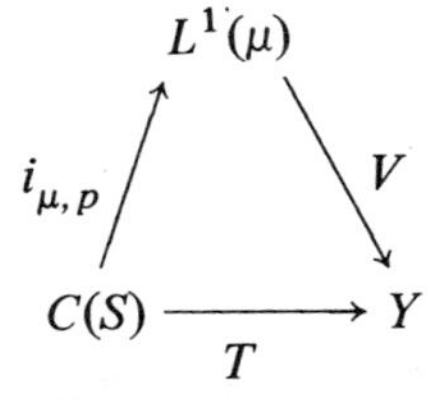

for some $\mu \in [C(S)]^*_+$ and a $V \in B(L^1(\mu), Y)$. Since V admits a representation (0.8), it takes weakly convergent sequences in $L^1(\mu)$ into norm convergence sequences in Y (cf. [**D-SI**, Chapter VI, §8, Theorem 14]). On the other hand, $i_{\mu,1}$ is a weakly compact operator. Hence $Vi_{\mu,1} = T$ is compact. □

Finally recall that if X is a finite dimensional Banach space and $T \in B(X, X)$, then the *trace of* T is defined by

$$\operatorname{tr}(T) = \sum_{j=1}^{\dim X} e_j^*(Te_j)$$

where $(e_j)_{1 \leqslant j \leqslant \dim X}$ is any basis for X and $(e_j^*)_{1 \leqslant j \leqslant \dim X}$ the sequence of coefficient functionals of this basis. The quantity $\operatorname{tr}(T)$ is independent of the choice of a basis. It satisfies the inequality

$$|\operatorname{tr}(T)| \leqslant n(T).$$

1. The F. and M. Riesz Theorem and Duals of the Disc Algebra

This section has an introductory character. We outline a proof of the classical M. and F. Riesz theorem and use it to describe the duals of the disc algebra A and the quotient $C(\partial D)/A$.

Let us regard the disc algebra A as a subspace of $C(\partial D)$. The annihilator

$$A^{\perp} = \left\{\mu \in [C(\partial D)]^*: \int_{\partial D} f\,d\mu = 0 \text{ for } f \in A\right\}$$

is completely described by the classical

THEOREM 1.1 (F. AND M. RIESZ). *Let m be Lebesgue measure on the unit circle $\partial D = \{z \in C: |z| = 1\}$. Then*

$$A^{\perp} = \{\mu \in [C(\partial D)]^* : \mu = h \cdot m \text{ for some } h \in H_0^1\}$$

where $H_0^1 = \{$the closure in $L^1 = L^1(m)$ of all the polynomials $\Sigma_{j=1}^n c_j z^j$ $(n = 1, 2, \ldots)\}$.

Briefly $A^{\perp} = H_0^1$ if we identify (via the Radon-Nikodym theorem) the space L^1 with the subspace $\{\mu \in [C(S)]^* : \mu \ll m\}$.

SKETCH OF THE PROOF. *Step* I. A closed set $F \subset \partial D$ with $m(F) = 0$ is a peak set of A, i.e., there is an $f_F \in A$ such that $f_F(z) = 1$ iff $z \in F$, $|f_F(z)| < 1$ for $z \notin F$.

We put

$$f(z) = 1 - \exp \frac{1}{2\pi}\int_0^{2\pi} \frac{z + e^{it}}{z - e^{it}} \log \omega(t)\,dt \quad \text{for } |z| < 1$$

where $\omega(\cdot)$ is a real 2π-periodic continuous function such that $0 \leqslant \omega \leqslant 1$, $\int_0^{2\pi} \log \omega(t)\,dt > -\infty$, $\omega(t) = 0$ iff $e^{it} \in F$, and ω is infinitely many times differentiable at each point t with $e^{it} \notin F$.

$$f_F(z) = \lim_{\rho \to 1} f(\rho z) \quad \text{for } z \in \partial D.$$

The essential technical difficulty is to show that the limit exists and that f is uniformly continuous for $|z| < 1$ (cf. [H, p.80], [St, p. 205] for details).

Step II. If $\mu \in A^{\perp}$, then $\mu \ll m$. Pick any closed $F \subset \partial D$ with $m(F) = 0$ and define f_F as in Step I. Then

$$0 = \int f_F^n\,d\mu \quad \text{for } n = 1, 2, \ldots .$$

Hence $\mu(F) = \lim_n \int f_F^n\,d\mu = 0$. Thus $\mu \ll m$, being a regular Borel measure which vanishes on every closed subset of ∂D of Lebesgue measure zero.

Step III. $A^\perp = H^0_1$. By Step II and the Radon-Nikodym theorem, if $\mu \in A^\perp$, then $\mu = h \cdot m$ for some $h \in L^1(m)$. Let $f(t) = h(e^{it})$. Then $\mu \in A^\perp$ yields

$$\hat{f}(-n) = \frac{1}{2\pi}\int_0^{2\pi} f(t)e^{int}\,dt = 0 \quad \text{for } n = 0, 1, 2, \ldots .$$

Thus, by the Fejer theorem, f is the limit in L^1 of polynomials of the form $\Sigma_{j=1}^k c_j e^{ijt}$. Equivalently $h \in H^1_0$. Thus $H^1_0 \supset A^\perp$. Obviously $z^j \in A^\perp$ for $j = 1, 2, \ldots$. Hence $H^1_0 \subset A^\perp$. □

The description of the dual of A. If Y is a subspace of a Banach space X, then the map which assigns to the coset $\{x^* + Y^\perp\}$ the restriction of x^* to Y is an isometric isomorphism from $X^*/Y^\perp$ onto Y^*. Thus, by the M. and F. Riesz theorem, A^* is isometrically isomorphic to $[C(\partial D)]^*/H^1_0$. By the Lebesgue decomposition theorem

(1.1) $$[C(\partial D)]^* = L^1 \oplus_1 V_{\text{sing}}.$$

(The equality in (1.1) and (1.2) below means "isometric isomorphism" and the symbol $X \oplus_1 Z$ means the l^1-sum of X and Z.) Thus we have

(1.2) $$A^* = L^1/H^1_0 \oplus_1 V_{\text{sing}}.$$

It follows from the definition of H^∞ that

$$(H^1_0)^\perp = \left\{f \in L^\infty = L^\infty(m) \colon \int fg\,dm = 0 \text{ for } g \in H^1_0\right\} = H^\infty.$$

Thus a standard duality argument gives $(L^1/H^1_0)^* = H^\infty$. Now (1.2) yields

(1.3) $$A^{**} = H^\infty \oplus_\infty V^*_{\text{sing}}.$$

(The symbol $X \oplus_\infty Z$ means the l^∞-sum of X and Z.) Observe that V_{sing} is an $L^1(\nu)$-space. Hence V^*_{sing} is a $C(K)$-space for an extremely disconnected compact Hausdorff space K.

Our next aim is to state some consequences of the F. and M. Riesz theorem and (1.2) in terms of isomorphic invariants of Banach spaces.

A subspace E of a $C(S)$-space is said to have a *small annihilator* if there is a $\nu \in [C(S)]^*_+$ such that $E^\perp \subset L^1(\nu)$. Obviously if $E^\perp$ is a norm separable subspace of $[C(S)]^*$, then E has a small annihilator. Conversely if S is metrizable and $E \subset C(S)$ has a small annihilator then $E^\perp$ is norm separable.

We will distinguish two properties of a Banach space X.

Property I. X is isomorphic to a space with a small annihilator.

Property II. X^* is a *separable distortion* of an $L^1(\nu)$, i.e., X^* is isomorphic to a product $M \oplus V$ where M is a separable space and V is an $L^1(\nu)$ space.

Observe the following: 1°. The disc algebra has both Properties I and II.

2°. If X is separable then Property I implies Property II.

3°. If $n \geqslant 2$ then n-dimensional Hilbert space is not isometrically isomorphic to any subspace of a $C(S)$-space with a small annihilator. However, every finite dimensional Banach space (being isomorphic to a $C(S)$-space) has Property I.

We end this section by stating a description of a predual and the dual of H^1 which is also a simple consequence of the F. and M. Riesz theorem and the following general fact. If Y is a subspace of a Banach space X then the dual of X/Y is naturally isometrically isomorphic to $Y^\perp$ and $(Y^\perp)^*$ is naturally isometrically isomorphic to X^{**}/Y^{**}. We put $A_0 = zA$ and $H_0^\infty = zH^\infty$.

THEOREM 1.2. (i) *The map* $f \longrightarrow x_f^*$ *where* $x_f^*(\{g + A_0\}) = \int_{\partial D} fg\, dm$ *for* $g \in C(\partial D)$ *is an isometric isomorphism between* H^1 *and the dual* $(C(\partial D)/A_0)^*$.

(ii) *The map* $\{g + H_0^\infty\} \longrightarrow \varphi_g^*$ *for* $g \in L^\infty$ *where* $\varphi_g^*(f) = \int_{\partial D} fg\, dm$ *for* $f \in H^1$ *is an isometric isomorphism from* L^∞/H_0^∞ *onto* $(H^1)^*$.

(iii) *The map* $\{f + A_0\} \longrightarrow \{f + H_0^\infty\}$ *for* $f \in C(\partial D)$ *can be identified with the canonical embedding of* $C(\partial D)/A_0$ *into its second dual identified with* L^∞/H_0^∞. □

2. Absolutely Summing Operators from the Disc Algebra

The main result of this section is Theorem 2.3 describing p-absolutely summing operators from the disc algebra. It is preceded by a description of p-absolutely summing operators from a subspace of a $C(S)$-space with a small annihilator (Theorem 2.2). The technique is based upon Bishop's generalized Rudin-Carleson theorem **[B1]**, **[Gl1]** which with the improvement of **[Gm2]**, **[P2]**, **[P3]**, **[M-P]** (cf. also **[Gm1]**, **[St]**) can be stated as follows.

THEOREM 2.1. *Let X be a subspace of a $C(S)$-space. Let F be a closed subset of S such that*

$$\mu(F_1) = 0 \quad \text{for every closed } F_1 \subset F \text{ and every } \mu \in X^{\perp}. \tag{2.1}$$

Then for every $u \in C(F)$, every $\epsilon > 0$ and every open $G \supset F$ there is an $f_u \in X$ such that

$$\begin{aligned} f_u(s) &= u(s) \quad \text{for } s \in F, \\ |f_u(s)| &< \epsilon \quad \text{for } s \notin G \text{ and } \|f\| = \|u\|. \end{aligned} \tag{2.2}$$

Moreover, if X is separable then the map $u \longrightarrow f_u$ from $C(F)$ into X can be chosen to be a linear isometry.

Let us observe that if X has a small annihilator in $C(S)$, say $X^{\perp} \subset L^1(\lambda)$ for some $\lambda \in [C(S)]^*_+$, then a closed $F \subset S$ satisfies (2.1) whenever $\lambda(F) = 0$.

Now we are ready to prove

THEOREM 2.2 (MITJAGIN-PELCZYNSKI). *Let X be a subspace of a $C(S)$-space with $X^{\perp} \subset L^1(\lambda)$ for some probability measure $\lambda \in [C(S)]^*$.*

Let $1 \leqslant p < \infty$ and let $T\colon X \longrightarrow E$ be a p-absolutely summing operator into a Banach space E.

Then there exists a strictly p-integral operator $V\colon X \longrightarrow E$ and a nonnegative $h \in L^1(\lambda)$ such that

$$\|(T - V)(f)\| \leqslant \left(\int_S |f(s)|^p h(s)\, d\lambda\right)^{1/p} \quad \text{for } f \in X, \tag{2.3}$$

$$(\pi_p(T))^p - (i_p(V))^p \geqslant \int_S h(s)\, d\lambda. \tag{2.4}$$

PROOF. Since T is p-absolutely summing, there exists a $\mu \in [C(S)]^*_+$ such that $\pi_p(T) = \|\mu\|^{1/p}$ and $\|Tf\| \leqslant (\int_S |f(s)|^p\, d\mu)^{1/p}$ for $f \in X$ (cf. §0.III). Hence there is a unique bounded linear operator $B\colon X_\mu \longrightarrow E$ such that $B(i_{\mu,p}(f)) = T(f)$ for $f \in X$. For a

$\tau \in [C(S)]^*_+$ we denote by $X_{\tau,p}$ the closure of X (more precisely, of $i_{\tau,p}(X)$) in $L^p(\tau)$. The Lebesgue decomposition theorem yields $\mu = \sigma + \nu$ with $\sigma \ll \lambda$, $\nu \perp \lambda$ and $\|\mu\| = \|\sigma\| + \|\nu\|$. We shall identify $L^p(\mu)$ with the l^p-direct sum $L^p(\sigma) \oplus_p L^p(\nu)$. The crucial point of the proof is to establish

(2.5) $$X_{\mu,p} = X_{\sigma,p} \oplus_p L^p(\nu).$$

Assume that (2.5) has been proved. We then define $V\colon X \longrightarrow E$ by $Bj_\nu Q_\nu i_{\mu,p}$ where Q_ν: $L^p(\sigma) \oplus_p L^p(\nu) \longrightarrow L^p(\nu)$ is the natural projection, $j_\nu\colon L^p(\nu) \longrightarrow L^p(\sigma) \oplus_p L^p(\nu)$ the natural embedding (i.e., $j_\nu(b) = (0, b)$ for $b \in L^p(\nu)$). By (2.5), $j_\nu Q_\nu i_{\mu,p}(X) \subset X_{\mu,p}$; thus V is well defined. Clearly $Q_\nu i_{\mu,p} = i_{\nu,p}$, and $\|Bj_\nu\| \leqslant \|B\|\,\|j_\nu\| \leqslant 1$. Thus V is strictly p-integral with $i_p(V) \leqslant \|\nu\|^{1/p}$. Next for each $f \in X$

$$\|(T - V)(f)\| = \|Bi_{\mu,p}(f) - Bj_\nu Q_\nu i_{\mu,p}(f)\|$$
$$\leqslant \|i_{\mu,p}(f) - j_\nu Q_\nu i_{\mu,p}(f)\|.$$

If $h \in L^p(\mu) = L^p(\sigma) \oplus_p L^p(\nu)$ then $h = j_\sigma Q_\sigma(h) + j_\nu Q_\nu(h)$ where $Q_\sigma : L^1(\mu) \longrightarrow L^1(\sigma)$ and $j_\sigma\colon L^1(\sigma) \longrightarrow L^1(\mu)$ are the natural projection and the natural embedding respectively. Thus for $h = i_{\mu,p}(f)$, we have $(f) - j_\nu Q_\nu i_{\mu,p}(f) = j_\sigma Q_\sigma i_{\mu,p}(f) = i_{\sigma,p}(f)$. It follows that

$$\|(T - V)(f)\| \leqslant \|i_{\sigma,p}(f)\| = \left(\int_S |f|^p \, d\sigma\right)^{1/p} \quad \text{for } f \in X.$$

Hence we get (2.3) with $h = d\sigma/d\lambda$. Finally

$$(\pi_p(T))^p - (i_p(V))^p \geqslant \|\mu\| - \|\nu\| = \|\sigma\| = \int_S h \, d\lambda$$

which proves (2.4).

PROOF OF (2.5). The inclusion $X_{\mu,p} \subset X_{\sigma,p} \oplus_p L^p(\nu)$ is trivial. For proving the reverse one, pick a z in $X_{\sigma,p} \oplus_p L^p(\nu)$, i.e., pick an $a \in X_{\sigma,p}$ and $b \in L^p(\nu)$ so that $z = a + b$ and

(2.6) $$\int_S |a + b|^p \, d\mu = \int_S |a|^p \, d\sigma + \int_S |b|^p \, d\nu.$$

Next fix $\epsilon > 0$ and pick an $x \in X$ so that $\int_S |a - x|^p \, d\sigma < \epsilon^p$. Since $b \in L^p(\nu)$, there is a finite $M > 1$ such that if $Z = \{s \in S\colon |b(s)|^p \leqslant M^p\}$, then

(2.7) $$\int_{S\setminus Z} |b|^p \, d\nu < \epsilon^p \quad \text{and} \quad \nu(S\setminus Z) \cdot M^p < \epsilon^p.$$

Combining (1.6), (1.7) with the Lusin theorem and the fact that ν, σ, λ are regular Borel measures with $\nu \perp \lambda$, and $\sigma \ll \lambda$ and using the fact that the measure $B \longrightarrow \int_B |b(s)|^p \, d\nu$ is absolutely continuous with respect to ν, we construct a compact $F \subset Z$ and an open $G \supset F$ such that

(2.8)
$$b \text{ restricted to } F \text{ is continuous, } \lambda(F) = 0,$$
$$\max\left\{\nu(S\setminus F),\ \sigma(G),\ \int_{Z\setminus F} |b|^p \, d\nu\right\} < \epsilon^p/M^p.$$

Now we apply Theorem 2.1 for $u \in C(F)$ defined by $u(s) = b(s) - x(s)$ for $s \in F$.

Let $f_u \in X$ satisfy (2.2). Then $\|f_u\| \leqslant \sup_{s \in F}|b(s)| + \|x\| \leqslant M + \|x\|$ because $F \subset Z$. Hence, we get

$$\left(\int_S |z - (x + f_u)|^p \, d\mu\right)^{1/p} \leqslant \left(\int_S |a - x - f_u|^p \, d\sigma\right)^{1/p} + \left(\int_S |b - x - f_u|^p \, d\nu\right)^{1/p}.$$

Estimating each of the terms separately we get

$$\begin{aligned}
\left(\int_S |a - x - f_u|^p \, d\sigma\right)^{1/p} &\leqslant \left(\int_S |a - x|^p \, d\sigma\right)^{1/p} + \left(\int_S |f_u|^p \, d\sigma\right)^{1/p} \\
&\leqslant \epsilon + \left(\int_{S \setminus G} |f_u|^p \, d\sigma\right)^{1/p} + \left(\int_G |f_u|^p \, d\sigma\right)^{1/p} \\
&\leqslant \epsilon + \sigma(S)^{1/p}\epsilon + (\sigma(G)\|f_u\|)^{1/p} \\
&\leqslant \epsilon(1 + \sigma(S)^{1/p} + 1 + \|x\|/M);
\end{aligned}$$

$$\begin{aligned}
\int_S |b - x - f_u|^p \, d\nu &= \left(\int_{S \setminus F} |b - x - f_u|^p \, d\nu\right)^{1/p} \\
&\leqslant \left(\int_{S \setminus F} |b|^p\right)^{1/p} + \left(\int_{S \setminus F} |x + f_u|^p \, d\nu\right)^{1/p} \\
&\leqslant \left(\int_{Z \setminus F} |b|^p \, d\nu\right)^{1/p} + \left(\int_{S \setminus Z} |b|^p \, d\nu\right)^{1/p} + [\nu(S \setminus F)]^{1/p}\|x + f_u\| \\
&\leqslant (c + \epsilon + \epsilon(2\|x\| + M))M^{-1} \\
&\leqslant \epsilon(2 + 2\|x\| + M)M^{-1}.
\end{aligned}$$

Letting ϵ tend to zero we infer that $z \in X_\mu$. □

COROLLARY 2.1. *Let X satisfy the assumption of Theorem 2.1. If E has the Radon-Nikodym property and $p = 1$ then the operator V satisfying* (2.3) *and* (2.4) *is nuclear with* $n(V) = i_1(V)$ *and therefore compact.*

PROOF. Use the fact that every strictly integral operator to a space with the Radon-Nikodym property is nuclear (cf. §0.III).

REMARK. The assumption of Corollary 2.1 is satisfied if E is either a Hilbert space or $E = H^1$, because H^1 being a separable dual (by Theorem 1.2) has the Radon-Nikodym property.

For the disc algebra Theorem 2.2 can be strengthened as follows:

THEOREM 2.3. *Let E be a Banach space, $1 \leqslant p < \infty$, and let $T: A \longrightarrow E$ be a p-absolutely summing operator. Then there is a strictly p-integral operator $V: A \longrightarrow E$ and an outer function $\varphi \in H^1$ such that*

$$\|(T - V)(f)\| \leqslant \left(\int |f|^p |\varphi| \, dm\right)^{1/p}, \tag{2.9}$$

$$|\varphi(z)| \geqslant 1 \quad \text{for } |z| < 1, \tag{2.10}$$

(2.11) $$1 + \pi_p(T)^p - i_p(V)^p \geqslant \int |\varphi|\, dm.$$

PROOF. Let V and $h \in L^1$ be defined to satisfy (2.3) and (2.4) (for $S = \partial D$, $\lambda = m$, $X = A$). Let φ be the outer function defined by

$$\varphi(z) = \exp \frac{1}{2\pi} \int_0^{2\pi} \frac{e^{it} + z}{e^{it} - z} \log[h(e^{it}) + 1]\, dt \quad \text{for } |z| < 1.$$

It is well known [**Du**, Chapter 2] that $\varphi \in H^1$ and

$$|\varphi(e^{it})| = \lim_{\rho \to 1} |\varphi(\rho e^{it})| = h(e^{it}) + 1 \quad \text{almost everywhere}$$

which implies (2.9) and (2.11).

Since $h \geqslant 0$, for $z = re^{it}$ with $r < 1$ we have

$$|\varphi(z)| = \exp \frac{1}{2\pi} \int_0^{2\pi} \operatorname{Re}\left(\frac{e^{it} + z}{e^{it} - z}\right) \cdot \log(1 + h(e^{it}))\, dt$$

$$= \exp \frac{1}{2\pi} \int_0^{2\pi} \frac{1 - r^2}{1 + r^2 - 2r\cos(t - \arg z)} \log(1 + h(e^{it}))\, dt$$

$$\geqslant 1,$$

which proves (2.10).

THEOREM 2.4 (MITJAGIN-PELCZYNSKI). *Let $1 < p < \infty$. Then every p-absolutely summing operator from A is strictly p-integral. More precisely, there is a constant C_p with*

(2.12) $$C_p \leqslant \frac{p^2}{p - 1} b$$

(where b is an absolute positive constant independent of p) such that $i_p(T) \leqslant C_p \pi_p(T)$ for every p-absolutely summing operator from A.

PROOF. Let φ be an outer function in H^1 satisfying (2.10). Let $i^A_{|\varphi|m,p}$ denote the restriction of the natural injection $i_{|\varphi|m,p} : C(\partial D) \longrightarrow L^p(|\varphi|m)$ to A regarded as an operator from A into $H^p_{|\varphi|m}$ where $H^p_{|\varphi|m}$ is the closure of $i_{|\varphi|m,p}(A)$ in $L^p(|\varphi|m)$. We first prove

(2.13) $i^A_{|\varphi|m,p}$ is a strictly p-integral operator with

$$i_p(i^A_{|\varphi|m,p}) \leqslant \|R\|_p \left(\int_{\partial D} |\varphi|\, dm\right)^{1/p}.$$

Here $\|R\|_p$ denotes the norm of the orthogonal (Riesz) projection from L^p onto H^p regarded as an operator on L^p.

Let us consider the diagram

(2.14)
$$\begin{array}{ccccccc} C(\partial D) & \xrightarrow{i_{|\varphi|m,p}} & L^p(|\varphi| \cdot m) & \xrightarrow{M_\psi} & L^p & \xrightarrow{R} & H^p \\ \uparrow j & & & & & & \downarrow M_{\psi^{-1}} \\ A & & & \xrightarrow{i^A_{|\varphi|m,p}} & & & H^p_{|\varphi| \cdot m} \end{array}$$

where j denotes the natural inclusion, ψ is the outer function defined by

$$\psi(z) = \exp\left(\frac{1}{p}\,\frac{1}{2\pi}\int_0^{2\pi} \frac{z+e^{it}}{z-e^{it}} \log \operatorname{Re} \varphi(t)\,dt\right) \quad \text{for } |z|<1,$$

and M_ψ and $M_{\psi^{-1}}$ the operators of multiplications by ψ and ψ^{-1} respectively, i.e.,

$$M_\psi(f) = \psi \cdot f \quad \text{and} \quad M_{\psi^{-1}}(f) = \psi^{-1} f.$$

It is known that $|\psi|^p = |\varphi|$ and $\psi \in H^p$ (because $\varphi \in H^1$). Hence M_ψ is an isometric isomorphism from $L^p(|\varphi|m)$ onto L^p. We have to check that

(2.15) $$M_\psi \text{ carries } H^p_{|\varphi|m} \text{ onto } H^p.$$

This will show that the diagram (2.14) is commutative and therefore $i^A_{|\varphi|m,p}$ is strictly p-integral.

To prove (2.15) observe that obviously $M_\psi(H^p_{|\varphi|m}) \subset H^p$; hence it suffices to show

(2.16) $$\text{if } g \in H^p, \text{ then } g/\psi \in H^p_{|\varphi|m}.$$

Let $\psi_r(z) = \psi(rz)$ for $|z| \leqslant 1$ and $0<r<1$. Clearly $\psi_r \in A$ and, by (2.10), $|\psi_r(z)| = |\varphi(rz)|^{1/p} \geqslant 1$ for $|z| \leqslant 1$. Observe that if $f \in A$ then $f/\psi \in H^p_{|\varphi|m}$ because $f/\psi_r \in A$ and

$$\overline{\lim_{r=1}} \int \left|\frac{f}{\psi} - \frac{f}{\psi_r}\right|^p |\varphi|\,dm \leqslant \|f\|_\infty \overline{\lim_{r=1}} \int \frac{|\psi_r - \psi|^p}{|\psi_r|^p}\,dm \leqslant \|f\|_\infty \overline{\lim_{r=1}} \int |\psi_r - \psi|^p\,dm = 0$$

(we use here that $\psi \in H^p$ (cf. [**Du**, Chapter 2])). Next given $g \in H^p$, there is a sequence (f_n) in A with $0 = \lim_n \int_{\partial D} |g - f_n|^p\,dm = \lim_n \int_{\partial D} |g/\psi - f_n/\psi|^p\,|\varphi|\,dm$. Since $f_n/\psi \in H^p_{|\varphi|m}$ for all n, this yields (2.16). The commutativity of the diagram (2.14) means that $i^A_{|\varphi|m,p} = M_{\psi^{-1}} R M_\psi i_{|\varphi|m,p} \cdot j$. Since $\|M_\psi\| = \|M_{\psi^{-1}}\| = \|j\| = 1$, we have $i_p(i^A_{|\varphi|m,p}) \leqslant \|R\|_p\, i_p(i_{|\varphi|m,p}) = \|R\|_p (\int |\varphi|\,dm)^{1/p}$. This completes the proof of (2.13).

Now, let E be an arbitrary Banach space, $1<p<\infty$, and let $T\colon A \longrightarrow E$ be a p-absolutely summing operator with $\pi_p(T) = 1$. By Theorem 2.3, there is a strictly p-integral operator $V\colon A \longrightarrow E$ and an outer $\varphi \in H^1$ satisfying the conditions (2.9)–(2.11). It follows from (2.9) that there is a unique bounded linear contraction $B\colon H^p_{|\varphi|m} \longrightarrow E$ such that $B(i_{|\varphi|m,p}(f)) = (T-V)(f)$ for $f \in A$, equivalently $B \cdot i^A_{|\varphi|m,p} = T - V$. Thus, by (2.13)

$$i_p(T-V) \leqslant \|B\|\, i_p(i^A_{|\varphi|p}) \leqslant \|R\|_p \left(\int |\varphi|\,dm\right)^{1/p}.$$

Since $\pi_p(T) = 1$, (2.11) yields $(\int |\varphi|\,dm)^{1/p} \leqslant 2^{1/p} < 2$. Hence, again using (2.11), $i_p(T) \leqslant i_p(T-V) + i_p(V) \leqslant 2\|R\|_p + 2 = C_p$.

This proves the last assertion of Theorem 2.4. Finally (2.12) follows from the estimation for $\|R\|_p$ (cf. §0.II). □

REMARK. Let us observe that the estimation for C_p given in (2.12) is the best possible in the following sense

$$\inf i_p(T) \geqslant ap^2/(p-1) \quad \text{for } 1<p<\infty$$

where a is an absolute constant independent of p and inf is extended over all operators from A with $\pi_p(T) = 1$. In fact for the natural embedding $i^A_{m,p}\colon A \longrightarrow H^p$ we have $\pi_p(i^A_{m,p}) = 1$ and

(2.17) $$i_p(i^A_{m,p}) = \|\mathcal{R}\|_p,$$

PROOF OF (2.17). Since $i^A_{m,p} = \mathcal{R}\, i_{m,p} j$ we have $\|\mathcal{R}\|_p \geqslant i_p(i^A_{m,p})$ (here $j\colon A \longrightarrow C(\partial D)$ denotes the natural inclusion). To prove the reverse inequality we use the standard averaging technique (cf. **[P10]** for a more general treatment). Pick $\mu \in [C(\partial D)]^*_+$ with $\|\mu\| = 1$ so that $i^A_{m,p} = B i_{\mu,p} j$ for some operator $B\colon L^p(\mu) \longrightarrow H^p$ with $\|B\| = i_p(i^A_{m,p})$. (The existence of B follows from the definition of p-integral operators (cf. §0.III) and the fact that H^p is reflexive for $1 < p < \infty$ and therefore every integral operator to H^p is strictly integral.) Let us set

$$(S_\alpha f)(z) = f(\alpha z) \quad \text{for } f \in L^1 \text{ and } |\alpha| = |z| = 1,$$

$$\widetilde{B}(f) = \int S_\alpha B S_{\alpha^{-1}} f\, m(d\alpha) \quad \text{for } f \in C(\partial D).$$

Then, we have, for every $f \in C(\partial D)$,

$$\begin{aligned}\|\widetilde{B}f\|_p^p &\leqslant \int \|S_\alpha B S_{\alpha^{-1}} f\|_p^p\, m(d\alpha) = \int \|B S_{\alpha^{-1}} f\|_p^p\, m(d\alpha)\\ &\leqslant \|B\| \int \left(\int |S_{\alpha^{-1}} f|^p\, d\mu \right) m(d\alpha) = \|B\| \int |f|^p\, dm.\end{aligned}$$

(The last equality follows from the observation that because of the uniqueness of the Haar measure $\int S_{\alpha^{-1}} h\, d\mu m(d\alpha) = \int h\, dm$ for every probability measure $\mu \in C(\partial D)^*$ and every non-negative $h \in L^1$.) Thus $\widetilde{B}$ extends to a bounded linear operator from L^p into H^p which coincides with $\mathcal{R}$ because $Bf = f$ for $f \in A$. Thus $\|\mathcal{R}\|_p \leqslant \|B\| = i_p(i^A_{m,p})$. □

3. Absolutely Summing Operators from the Disc Algebra into Hilbert Space

The operator $f \longrightarrow (\hat{f}(2^n))$ is an example of an absolutely summing map from A onto l^2. The existence of this phenomenon leads to a new proof of Grothendieck's theorem that every bounded operator from l^1 into l^2 is absolutely summing. It also emphasizes the difference between the disc algebra and $C(S)$-spaces, because every absolutely summing operator from a $C(S)$-space into l^2 is nuclear, hence compact. The main result of this section—Theorem 3.2—gives a description of absolutely summing operators with a closed range from A into l^2. At the end of this section we discuss the problem whether every bounded operator from A into l^2 is 2-absolutely summing.

The following concept plays an important role throughout this section.

DEFINITION 3.1. An orthogonal projection $P: H^2 \longrightarrow H^2$ is a *Paley projection* if there is a $K > 0$ such that

$$\|Pf\|_2 \leqslant K\|f\|_1 \quad \text{for } f \in H^2. \tag{3.1}$$

EXAMPLE 3.1. Let (n_k) be a sequence of nonnegative integers such that

$$\varliminf_k n_{k+1}/n_k > 1.$$

Then the orthogonal projection $Pf = \Sigma_k \hat{f}(n_k) z^{n_k}$, $f \in H^2$ (where $\hat{f}(n) = \int_{\partial D} f(z) z^{-n}\, m(dz)$ for $n = 0, 1, \ldots$) is a Paley projection. This fact is simply a restatement of a classical result of Paley (cf. **[Z]**, **[Du**, Chapter 6**]**, **[Pa1]**).

If $P: H^2 \longrightarrow H^2$ is an operator then P_A denotes the restriction of P to A regarded as an operator from the disc algebra A into H^2, formally $P_A = Pi^A_{m,2}$. Our next result shows that every Paley projection induces an absolutely summing surjection from A onto a Hilbert space. Precisely we have

PROPOSITION 3.1. *If P is a Paley projection, then P_A is an absolutely summing surjection from A onto $P(H^2)$.*

PROOF. Clearly (3.1) implies (in fact is equivalent to)

$$\|P_A(f)\| \leqslant K \int_{\partial D} |f|\, dm \quad \text{for } f \in A. \tag{3.2}$$

Hence P_A is absolutely summing and $\pi_1(P_A) \leqslant K$.

To prove that P_A is onto $P(H^2)$ we show that P_A^*: $[P_A(H^2)]^* \longrightarrow A^*$ is an isomorphic embedding. Clearly we may identify the Hilbert space $[P_A(H^2)]^*$ with $P(H^2)$ and the adjoint P_A^* with the operator which assigns to each $f \in P(H^2)$ the functional x_f^* defined by

$$x_f^*(g) = \int_{\partial D} f(-z)g(z)m(dz) \quad \text{for } g \in A.$$

Here we used the fact that the projection P is orthogonal. By (1.2), the functional $x_f^* \in A^* = L^1/H_0' \times V_{\text{sing}}$ can be identified with the coset $\{f + H_0^1\}$ or, more precisely, with the pair $(\{f + H_0^1\}, 0)$. Therefore to prove that P_A^* is an isomorphic embedding it suffices to show that there is $K_1 > 0$ such that

$$\|P_A^*(f)\| = \inf_{h \in H_0^1} \int_{\partial D} |f(-z) + h(z)|\, m(dz) \geqslant K_1 \|f\|_2 \quad \text{for } f \in P(H^2). \tag{3.3}$$

By the Kolmogorov theorem the "orthogonal projection" annihilating H_0^1 is a bounded linear operator from L^1 into $L^{1/2}$ (cf. §0.II). Hence there is a $c > 0$ such that

$$\int_{\partial D} |f(-z) + h(z)|\, m(dz) \geqslant c\left(\int |f|^{1/2}\, dm\right)^2 \qquad (f \in P(H^2),\ h \in H_0^1). \tag{3.4}$$

Next, combining (3.1) with the Schwarz inequality applied to the product $|f|^{1/4}|f|^{3/4} = |f|$, we get

$$\begin{aligned}\left(\int |f|^2\, dm\right)^{1/2} &\leqslant K \int |f|\, dm \\ &\leqslant K\left(\int |f|^{1/2}\, dm\right)^{1/2}\left(\int |f|^{3/2}\, dm\right)^{2/3 \cdot 3/4} \\ &\leqslant K\left(\int |f|^{1/2}\, dm\right)^{1/2}\left(\int |f|^2\, dm\right)^{1/2 \cdot 3/4}.\end{aligned}$$

Hence

$$\left(\int |f|^2\, dm\right)^{1/2} \leqslant K\left(\int |f|^{1/2}\, dm\right)^2. \tag{3.5}$$

Combining (3.4) with (3.5) we get (3.3) with $K_1 = cK^{-4}$. □

COROLLARY 3.1. *There exists a map from the disc algebra onto l^2 which is absolutely summing.*

PROOF. Let (n_k) be a sequence of positive integers with $\underline{\lim}_k n_{k+1}/n_k > 1$. Then the operator $T\colon A \longrightarrow l_2$ defined by $Tf = (\hat{f}(n_k))$ for $f \in A$ has the desired properties. This follows immediately upon combining Example 3.1 with Proposition 3.1.

Corollary 3.1 has a surprising application.

THEOREM 3.1 (GROTHENDIECK [Gr1]). *Every bounded linear operator from l^1 into l^2 is absolutely summing.*

PROOF. Let $S\colon l^1 \longrightarrow l^2$ be any bounded linear operator and let $T\colon A \longrightarrow l^2$ be an absolutely summing surjection. By the open mapping principle, there is a constant $K_1 > 0$ such that given $x \in l^2$ there is an $f \in A$ with $\|f\| \leqslant K_1\|x\|$ and $Tf = x$. Hence if e_j is the jth unit vector of l_1, then there is an $f_j \in A$ such that $Tf_j = Se_j$ and $\|f_j\| \leqslant K_1\|S\|$. Now we define the lifting $\widetilde{S}\colon l^1 \longrightarrow A$ by $\widetilde{S}((t_j)) = \Sigma t_j f_j$ for $(t_j) \in l^1$. Then $S = T\widetilde{S}$ and $\|\widetilde{S}\| \leqslant K_1\|S\|$. Hence $\pi_1(S) \leqslant K_1\pi_1(T)\|S\|$. □

Our next result shows that up to a Banach space automorphism of a Hilbert space and of the disc algebra every absolutely summing surjection from A into a Hilbert space is a nuclear perturbation of a Paley projection. Instead of discussing "surjection onto Hilbert spaces" it will be more convenient for us to work with operators into l^2 with a closed range. If φ is a function in A with $\varphi(z) \neq 0$ for $|z| < 1$ then the operator M_φ defined by $M_\varphi(f) = \varphi f$ for $f \in A$ is called a *multiplication automorphism of* A.

THEOREM 3.2. *Let* $T: A \longrightarrow l^2$ *be an absolutely summing operator with a closed range. Then there is a nuclear operator* $V: A \longrightarrow l^2$ *such that for every* $\epsilon > 0$ *there exist a multiplication automorphism* $M_\varphi: A \longrightarrow A$, *a Paley projection* $P: H^2 \longrightarrow H^2$ *and an invertible operator* $U: P(H^2) \longrightarrow T(A)$ *such that* $\pi_1(T - (UP_A M_\varphi + V)) < \epsilon$.

PROOF. By Corollary 2.1, the remark after Corollary 2.1 and Theorem 2.2, there exists a nuclear operator $V: A \longrightarrow l^2$ and an outer function φ satisfying the conditions (2.9)–(2.11) for $p = 1$. Let $T_1 = T - V$. Then the inequality (2.9) can be rewritten

$$\|T_1 f\| \leqslant \int |f|\,|\varphi|\,dm \quad \text{for } f \in A.$$

Hence there exists a unique linear operator $Q: H^1_{|\varphi|m} \longrightarrow l^2$ with $\|Q\| \leqslant 1$ such that $T_1 = Qi^A_{|\varphi|m,1}$. Next fix r with $0 < r < 1$ and define $Q_r: H^1_{|\varphi_r|m} \longrightarrow l^2$ by

$$Q_r(h) = Q(h \cdot \varphi_r/\varphi) \quad \text{for } h \in H^1_{|\varphi_r|m} \tag{3.6}$$

where $\varphi_r(z) = \varphi(rz)$ for $|z| \leqslant 1$. To show that Q_r is "well defined" we have to check that if $h \in H^1_{|\varphi_r|m}$ then $h\varphi_r/\varphi \in H^1_{|\varphi|m}$. To this end fix $\delta > 0$ and pick $f \in A$ so that

$$\frac{\delta}{2} > \int |h - f|\,|\varphi_r|\,dm = \int |h\varphi_r - f\varphi_r|\,dm = \int \left| h\frac{\varphi_r}{\varphi} - f\frac{\varphi_r}{\varphi} \right| |\varphi|\,dm.$$

In view of **[Du**, Chapter 2] there exists an $R < 1$ so close to 1 that

$$\int |\varphi - \varphi_R|\,dm < \delta(2\|f\varphi_r\|_\infty + 1)^{-1}.$$

Then, by (2.10),

$$\int \left| f\frac{\varphi_r}{\varphi} - f\frac{\varphi_r}{\varphi_R} \right| |\varphi|\,dm \leqslant \|f\varphi_r\|_\infty \int |\varphi - \varphi_R|\,dm < \frac{\delta}{2}.$$

Hence

$$\int \left| h\frac{\varphi_r}{\varphi} - f\frac{\varphi_r}{\varphi_R} \right| |\varphi|\,dm < \delta.$$

Since $f\varphi_r/\varphi_R \in A$, the last inequality yields that $h\varphi_r/\varphi \in H^1_{|\varphi|m}$. Hence Q_r is well defined. Now put $T_r = Q_r i^A_{|\varphi_r|m,1}: A \longrightarrow l^2$. Then, by (3.6), for $f \in A$,

$$\|(T_1 - T_r)(f)\| = \left\| Q\left(f - f\frac{\varphi_r}{\varphi}\right) \right\| \leqslant \int \left| f - f\frac{\varphi_r}{\varphi} \right| |\varphi|\,dm$$
$$\leqslant \int |f|\,|\varphi - \varphi_r|\,dm.$$

Thus in view of **[Du**, Chapter 2],

$$\lim_{r=1} \pi_1(T_1 - T_r) = \lim_{r=1} \int |\varphi - \varphi_r| \, dm = 0. \tag{3.7}$$

It follows from (2.10) that for $0 < r < 1$ the operator M_{φ_r} is a multiplication automorphism. Hence, by the definition of T_r, we have

$$\begin{aligned} \|T_r M_{\varphi_r^{-1}}(f)\| &= \|Q(M_{\varphi_r^{-1}}(f) \cdot \varphi_r/\varphi\| \\ &= \|Q(f\varphi^{-1})\| \leqslant \int |f| \, dm \leqslant \left(\int |f|^2 \, dm\right)^{1/2}. \end{aligned} \tag{3.8}$$

Hence there exists a linear contraction, say $S_r \colon H^2 \longrightarrow l^2$, such that $S_r(f) = T_r M_{\varphi_r^{-1}}(f)$ for $f \in A$. Since A is dense in H^2 in the norm $\| \cdot \|_2$, S_r is uniquely determined and closure $S_r(H^2) =$ closure $(T_r M_{\varphi_r^{-1}}(A))$. Let $P = P^{(r)} \colon H^2 \longrightarrow H^2$ be the orthogonal projection with $\ker P = \ker S_r$. Then there exists a one-to-one bounded linear operator $U_r = U \colon P(H^2) = l^2$ such that $S_r = UP$. Now assume that U has a bounded inverse, i.e., there exists a bounded linear operator $V \colon U(P(H^2)) \longrightarrow P(H^2)$ such that VU is the identity on $P(H^2)$. Then, by (3.8) for $f \in A$,

$$\begin{aligned} \|P(f)\| &\leqslant \|V\| \, \|UP(f)\| = \|V\| \, \|T_r M_{\varphi_r^{-1}}(f)\| \\ &\leqslant \|V\| \int |f| \, dm. \end{aligned}$$

Thus $\|P(f)\| \leqslant \|V\| \int |f| \, dm$ for $f \in H^2$ (because for every $f \in H^2$ there is a sequence (f_n) in A such that $\lim_n \|f_n - f\|_2 = \lim_n \|f_n - f\|_1 = 0$). Hence, if U has a bounded inverse, then P is a Paley projection and

$$T_r = T_r M_{\varphi_r^{-1}} M_{\varphi_r} = UP_A M_{\varphi_r}$$

and given $\epsilon > 0$,

$$\pi_1(T - (V + UP_A M_{\varphi_r})) = \pi_1(T_1 - T_r) < \epsilon$$

for r sufficiently close to 1 (by (3.7)).

We complete the proof by showing that there is an r_0 with $0 < r_0 < 1$ such that if $r_0 < r < 1$, then $U = U_r$ does have a bounded inverse. To this end observe that $T_1 = T - V$ has a closed range because T does and V being nuclear is compact. Hence there is an $\epsilon > 0$ such that if $\widetilde{T} \colon A \longrightarrow l^2$ is an operator with $\|T_1 - \widetilde{T}\| < \epsilon$, then $\widetilde{T}$ has a closed range. Thus, by (3.7), there is an r_0 with $0 < r_0 < 1$ such that if $r_0 < r < 1$, then T_r has a closed range. Therefore, $T_r M_{\varphi_r^{-1}}$ also has a closed range and so does S_r. This implies that U has a closed range because $S_r = UP$ and P is a projection onto the domain of U. Thus U being one-to-one has a bounded inverse. □

The results of this section show that absolutely summing operators from the disc algebra into a Hilbert space are in general quite different in nature than absolutely summing operators from a $C(S)$-space into a Hilbert space. The situation seems to be different in the case of 2-absolutely summing operators.

Problem 3.1. Is it true that

(i) every bounded linear operator from A into l^2 is 2-absolutely summing?

Let us observe that (i) is equivalent to each of the following properties:

(ii) every operator from l^2 into l^2 which factors through A is Hilbert-Schmidt;

(iii) every bounded linear operator from A into l_2 extends to a bounded linear operator from $C(\partial D)$ into l^2;

(iv) every operator from L^1/H_0^1 into l^2 is 2-absolutely summing;

(v) every operator from l^2 into l^2 which factors through L^1/H_0^1 is Hilbert-Schmidt;

(vi) for every linear operator $T\colon L^1/H_0^1 \longrightarrow l^2$ there exists an operator $\widetilde{T}\colon L^1 \longrightarrow l^2$ such that $Tq = \widetilde{T}$ where $q\colon L^1 \longrightarrow L^1/H_0^1$ denotes the quotient map.

The equivalences (i) $\Longleftrightarrow$ (ii) and (iv) $\Longleftrightarrow$ (v) follow immediately from the definition of a 2-absolutely summing map. The equivalence of (i) and (iii) follows from the extension property of 2-absolutely summing operators and the fact that every bounded linear operator from a $C(S)$ space into l^2 is 2-absolutely summing (cf. **[Gr1]**, **[L-P1]**). The equivalences (ii) $\Longleftrightarrow$ (v) and (iii) $\Longleftrightarrow$ (vi) follow by a standard duality argument (we apply formula (1.2) and use the fact that every bounded linear operator from L^1 (resp. $C(\partial D)$) into l^2 is 2-summing.

Notes and remarks to §§2 *and* 3. The results of §2 are due to Mitjagin and Pelczynski. Theorem 2.2 generalizes Proposition 1 of **[Mt-P]** where the case $p = 1$ is considered. The proofs of Theorems 2.3 and 2.4 have not been published previously.

The following is open (the answer is "yes" if H^∞ has the bounded approximation property):

Problem 3.2. Is every p-absolutely summing operator from H^∞ into an arbitrary Banach space p-integral?

The results of §3 are due to Pelczynski and Wojtaszczyk. The fact that if $\underline{\lim}_k (n_{k+1}/n_k) > 1$ then the operator $f \longrightarrow (\hat{f}(n_k))$ maps the disc algebra onto l^2 is not new (cf., e.g., **[Four1]**, **[Four2]** and **[Vin1]**).

4. The Nonexistence of Local Unconditional Structure for the Disc Algebra and for its Duals

In this section we shall show that the disc algebra A as a Banach space is in a certain sense very different from a $C(S)$-space. A is not isomorphic to a quotient of any $C(S)$-space, nor is it complemented in any Banach space with a local unconditional structure. Most of the argument which leads to the above results is based upon the observation that the natural embedding $i^A_{m,1}: A \longrightarrow H^1$ does not factor through any L^1-space. To illustrate our approach we begin with the following simple

PROPOSITION 4.1. *The disc algebra and H^∞ are not isomorphic to quotients of $C(S)$-spaces.*

PROOF. Let us consider the following property of a Banach space X

(4.1) there is a noncompact absolutely summing operator from X into a separable dual Banach space.

Note that 1°. If X satisfies (4.1), then so does Y whenever there is an operator from Y onto X. (*Hint*. Use the Banach open mapping principle.)

2°. A and H^∞ do have (4.1); the desired operator is $i^A_{m,1}: A \longrightarrow H^1$, resp., $i^{H^\infty}_{m,1}: H^\infty \longrightarrow H^1$ (H^1 is a separable dual by Theorem 1.2).

3°. Every $C(S)$-space fails to have (4.1). Every absolutely summing operator from a $C(S)$-space to a separable dual is nuclear and therefore compact (cf. §0.III). □

COROLLARY 4.1. *Neither the disc algebra nor H^∞ is isomorphic to a complemented subspace of a $C(S)$-space (Pelczynski* **[P2]**, *Rosenthal* **[R1]**); *in particular A is uncomplemented in $C(\partial D)$ and H^∞ is uncomplemented in L^∞ (Rudin* **[Ru1]**).

We shall strengthen Proposition 4.1 in various directions. To state our results we recall some concepts and facts.

An operator $T: X \longrightarrow Y$ is *L^1-factorable* if there is an $L^1(\nu)$-space and operators $U: X \longrightarrow L^1(\nu)$ and $V: L^1(\nu) \longrightarrow Y$ with $VU = T$. A Banach space X is said to have *G-L l.u.st* (= Gordon-Lewis local unconditional structure) if there is a $K > 0$ and a function $E \longrightarrow (U_E, V_E, F_E)$ which assigns to every finite dimensional subspace E of X a finite dimensional space F_E with a basis (f_j) satisfying the condition $\|\Sigma t_j f_j\| = \|\Sigma |t_j| f_j\|$ for all scalars $t_1, t_2, \ldots, t_n$, and operators $U_E: E \longrightarrow F_E$, $V_E: F_E \longrightarrow X$ such that $V_E U_E(e) = e$ for $e \in E$ and $\|V_E\| \|U_E\| < K$. X is said to have l.u.st if moreover the above U_E and V_E are isomorphisms from E onto F_E and from F_E onto E respectively **[D-P-R]**.

REMARKS. R.I. If T is L^1-factorable so is T^{**}. The converse is true assuming that the range of T is isomorphic to a complemented subspace of a dual Banach space **[G-L]**, **[F-J-T]**.

R.II. X has G-L l.u.st iff X^{**} is isomorphic to a complemented subspace of a (complex) Banach lattice. Hence if X has G-L l.u.st so do X^* and complemented subspaces of X **[F-J-T]**.

R.III. If a Banach lattice Z does not contain an isomorph of c_0 then Z is complemented in Z^{**} (cf. **[Sch**, Theorem 2.10 and Proposition 5.15**]**).

The following important result due to Gordon and Lewis **[G-L]** links Banach lattices and spaces with l.u.st with absolutely summing operators.

THEOREM 4.1. *Let X have G-L l.u.st. Then every absolutely summing operator from X into a dual Banach space is L^1-factorable.*

COROLLARY 4.2. *If X is a domain of a non-L^1-factorable absolutely summing operator whose range is a dual Banach space then X and all the duals of X do not have G-L l.u.st.*

PROOF. The second adjoint of an absolutely summing operator is absolutely summing. Combine this fact with Theorem 4.1, R.I and R.II. □

Now we are ready to state the main result of this section.

THEOREM 4.2. (i) *Every L^1-factorable operator from A into H^1 is compact.*

(ii) *A and all the duals of A do not have G-L l.u.st.*

(iii) *If either A or H^∞ is a quotient of a Banach space Y with G-L l.u.st, then Y contains a complemented isomorph of l^1.*

(iv) *If L^1/H_0^1 is a subspace of a Banach space Z with G-L l.u.st, then Z contains l_n^∞ uniformly, i.e., given any positive integer n and $\epsilon > 0$ there is isomorphic embedding U_n: $l_n^\infty \longrightarrow Z$ with $\|U_n\| \|U_n^{-1}\| < 1 + \epsilon$.*

(v) *If L^1/H_0^1 is isomorphic to a subspace of a Banach lattice Z then Z contains an isomorph of c_0.*

PROOF. (i) Let $T = VU$: $A \longrightarrow H^1$ with V: $L^1(\nu) \longrightarrow H^1$ and U: $A \longrightarrow L^1(\nu)$. Since H^1 is a separable dual, V takes weakly compact sets in $L^1(\nu)$ into a compact set in H^1 (cf. **[D-SI**, Chapter VI, 8**]**). To complete the proof it is enough to show that U is weakly compact or equivalently that U^*: $[L^1(\nu)]^* \longrightarrow L^1/H_0^1 \times V_{\text{sing}}$ has the same property. Recall that every operator from a $C(S)$-space into a Banach space which does not contain isomorphs of c_0 is weakly compact **[P1]**. Thus every operator from $[L^1(\nu)]^*$ into L^1/H_0 is weakly compact. L^1/H_0^1 does not contain isomorphs of c_0, because L^1/H_0^1 is separable and complemented in a dual Banach space A^*, isomorphs of c_0 are complemented in every separable Banach space in which they are embedded (Sobczyk **[S]**, cf. also **[Ve]**), and isomorphs of c_0 are never complemented in dual Banach spaces **[B-P]**. Since V_{sing} is an $L^1(\mu)$-space every operator from a $C(S)$-space $[L^1(\nu)]^*$ into V_{sing} is weakly compact. Hence every operator from $[L^1(\nu)]^*$ into the product $L^1/H_0^1 \times V_{\text{sing}}$ is weakly compact.

(ii) The natural embedding $i_{m,1}^A$: $A \longrightarrow H^1$ is absolutely summing but not compact.

Hence by (i), $i^A_{m,1}$ is not L^1-factorable. Now we use Corollary 4.2.

(iii) Let $T = i^{H^\infty}_{m,1}q\colon Y \longrightarrow H^1$ where $q\colon Y \longrightarrow H^\infty$ is the quotient map. Assume that T is L^1-factorable. (By Theorem 4.1, this is the case where Y has G-L l.u.st.) Consider the diagram

$$\begin{array}{ccccc} Y & \xrightarrow{\;q\;} & H^\infty & \xrightarrow{\;i^{H^\infty}_{m,1}\;} & H^1 \\ & {\scriptstyle U}\searrow & & \nearrow{\scriptstyle V} & \\ & & L^1(\nu) & & \end{array}$$

By the open mapping principle, there is a $c > 0$ and y_n in Y with $\|y_n\| < c$ and $Ty_n = z^n$ for $n = 1, 2, \ldots$. Clearly $\|z^n - z^m\|_{H^1} \geqslant 1$ for $n \neq m$. Hence the sequence $(Uy_n) \subset L^1(\nu)$ does not contain weak Cauchy subsequences, because $V\colon L^1(\nu) \longrightarrow H^1$ takes weak Cauchy sequences into norm Cauchy sequences **[D-SI**, Chapter VI.8**]**. (H^1 is a separable dual.) Therefore, by a result of **[K-P]** and **[P-R]**, there is an infinite sequence (Uy_{n_k}) which is equivalent to the unit vectors of l^1 such that there is a projection P from $L^1(\nu)$ onto the closed linear span of the Uy_{n_k}'s. Hence PU maps Y onto l^1, which yields the desired conclusion for H^∞ because l^1 has the lifting property. The proof for A is the same.

(iv) If $L^1/H^1_0 \subset Z$ and Z has G-L l.u.st, then $H^\infty = (L^1/H^1_0)^*$ is a quotient of Z^*. By R.II, Z^* has G-L l.u.st. Hence, by (iii), Z^* contains a complemented isomorph of l^1 which implies that Z contains l^∞_n uniformly **[G-J]**.

(v) Assume to the contrary that there is an isomorphism $j\colon L^1/H^1_0 \longrightarrow Z$ where Z is a Banach lattice which does not contain isomorphs of c_0. By R.III, there is a projection $P\colon Z^{**} \xrightarrow[\text{onto}]{} Z$. Let us consider the diagram

$$\text{(4.2)}\qquad \begin{array}{ccccc} C/A_0 & \xrightarrow{\;i_*\;} & L^1/H^1_0 & \xrightarrow{\;j\;} & Z \\ & {\scriptstyle V^*_*}\searrow & & & \uparrow{\scriptstyle P} \\ & & C(S) & \xrightarrow{\;U^*\;} & Z^{**} \end{array}$$

where $i_*\colon C/A_0 \longrightarrow L^1/H^1_0$ is defined by $i_*(\{f + A_0\}) = \{f + H^1_0\}$. Clearly $(i_*)^* = i^{H^\infty}_{m,1}$. Thus the assumption that Z has G-L l.u.st combined with R.II and Theorem 4.1 yields that $(ji_*)^*\colon Z^* \longrightarrow H^1$ factors through an $L^1(\nu)$ space. Let $(ji_*)^* = VU$ be the factorization. Then $(ji_*)^{**} = U^*V^*\colon H^\infty \longrightarrow Z^{**}$ factors through a $C(S)$-space $(L^1(\nu))^*$. Let V^*_* denote the restriction of $V^*\colon L^\infty/H^\infty_0 \longrightarrow C(S)$ into a subspace C/A_0 identified with its canonical image in L^∞/H^∞_0 (cf. Theorem 1.2(iii)). It is easy to check that the diagram (4.2) commutes, i.e., $ji_* = PU^*V^*_*$. Since Z does not contain c_0, PU^* is weakly compact **[P1]** and therefore PU^* takes weak Cauchy sequences in $C(S)$ into norm Cauchy sequences in Z **[D-SI**, Chapter VI**]**. Since the space C/A_0 has a separable dual (namely, H^1), every bounded sequence in C/A_0 contains a weak Cauchy subsequence. This implies that $PU^*V^*_*$ is compact. On the other hand, ji_* is not compact because j is an isomorphic embedding and i_* is not compact (we have $\|i_*(\{z^{-n} + A_0\} - \{z^{-m} + A_0\})\|_{L^1/H^1_0} \geqslant 1$ for $n \neq m$). Thus $PU^*V^*_* \neq ji_*$, a contradiction. □

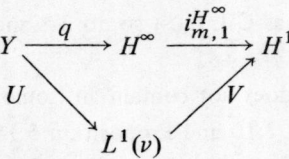

REMARK. The argument of Theorem 4.2(i) shows, in fact, the following:
Let X be a Banach space whose dual is a separable distortion of $L^1(\nu)$ *(cf.* §I). *Then*

(a) *Every operator from X into an* $L^1(\nu)$ *is weakly compact.*

(b) *Every* L^1*-factorable operator from X into a separable dual Banach space is compact.*

5. Application to Uniform Algebras

In this section we extend results of §4 and some results of §3 to uniform algebras with nontrivial Gleason parts, and to uniform algebras with a separable annihilator as well. We close by discussing Glicksberg's conjecture on uniform algebras on S complemented in $C(S)$ and related problems.

Recall that a *uniform algebra* on a (compact Hausdorff) space S is a closed subalgebra which separates the points of S and contains the constant functions.

We begin with the case of a uniform algebra whose annihilator is norm separable. Our first proposition combines a result of **[P4]** with a recent observation due to Wojtaszczyk **[W1]**.

For $\mu \in X^{\perp} = \{\nu \in [C(S)]^*: \int x\, d\nu = 0 \text{ for } x \in X\}$ we define $T_{\mu}: X \longrightarrow L^1(|\mu|)$ by $T_{\mu}(x) = xg$ for $x \in X$, where $g = d\mu/d|\mu|$ is the Radon-Nikodym derivative of μ with respect to $|\mu|$.

PROPOSITION 5.1. *Let X be a uniform algebra on S with a norm separable annihilator $X^{\perp} \subset [C(S)]^*$. Then the following conditions are equivalent*:

(i) $X = C(S)$.

(ii) *If $\mu \in X^{\perp}$, then μ is purely atomic.*

(iii) *If $\mu \in X^{\perp}$, then T_{μ} is compact.*

PROOF. (i) ⇒ (iii) obvious.

(iii) ⇒ (ii) (Wojtaszczyk). Assume that there is a non-purely-atomic $\mu \in X^{\perp}$, so that T_{μ} is compact. Then $T_{\mu}^*: L^{\infty}(|\mu|) \longrightarrow X^*$ is also compact. Since μ is not purely atomic, there is a sequence (r_n) in $L^{\infty}(|\mu|)$ which has the same distribution as the sequence of Rademacher functions in $L^1[0, 1]$. Observe that

(5.1) $$fr_n \xrightarrow[w^*]{} 0 \quad \text{as } n \longrightarrow \infty \text{ for every } f \in L^{\infty}(|\mu|)$$

(here "$\xrightarrow[w^*]{}$" denotes convergence in the $\sigma(L^{\infty}(|\mu|), L^1(|\mu|))$-topology). Since T_{μ}^* is weak-star continuous and compact, (5.1) yields

(5.2) $$\lim_n \|T_{\mu}^*(fr_n)\|_{X^*} = 0 \quad \text{for every } f \in L^{\infty}(|\mu|).$$

Using (5.2) we define inductively an increasing sequence of the indices $n(1) < n(2) < \cdots$ so that

(5.3) $$\|T_{\mu}^*(w_{j_1, j_2, \ldots, j_s})\|_{X^*} < 1/3^k \quad \text{for } j_s = k \ (k = 1, 2, \ldots)$$

where

$$w_{j_1,j_2,\ldots,j_s} = \prod_{k=1}^{s} r_{n(j_k)}$$

for any finite increasing sequence of positive integers $(j_k)_{1\le k\le s}$. Since X^* can be identified with $[C(S)]^*/X^\perp$, it follows from (5.3) that for every $w_{j_1,j_2,\ldots,j_s}$ there is a $v_{j_1,j_2,\ldots,j_s} \in X^\perp$ such that

$$(5.4) \qquad \|w_{j_1,\ldots,j_s} g|\mu| - v_{j_1,\ldots,j_s}\|_{[C(S)]^*} < 3^{-j_s},$$

(because $T^*_\mu(w_{j_1},\ldots,_{j_s}) = \{w_{j_1},\ldots,_{j_s} g|\mu| + X^\perp\}$). Let E denote the closed linear subspace of $L^1(|\mu|)$ spanned by all the "like-Walsh" functions $w_{j_1,j_2,\ldots,j_s}$. It is easy to see that E is isometrically isomorphic to $L^1[0, 1]$, by the isometry which assigns to the Walsh functions (in the Paley order) the functions $(w_{n_1,\ldots,n_k})$. Since $g = d\mu/d|\mu|$ is a unimodular function, the sequence $(gw_{n_1,\ldots,n_k})$ is also equivalent in $L^1(|\mu|)$ to the Walsh-Paley sequence in $L^1[0, 1]$ and the subspace gE is isometric to $L^1[0, 1]$. Now (5.4) yields that the perturbed sequence $(v_{j_1,j_2,\ldots,j_s})$ is also equivalent to the Walsh-Paley sequence in $L^1[0, 1]$, i.e., the map $w_{j_1,\ldots,j_s} \longrightarrow v_{j_1,\ldots,j_s}$ extends to an isomorphic embedding of E into $X^\perp$ (because the subspaces $E_k = \operatorname{span}\{w_{j_1,j_2,\ldots,j_s}$ with $j_s = k\}$ form a Schauder decomposition of E and $\dim E_k = 2^{k-1}$). Hence $X^\perp$ contains a subspace isomorphic to $L^1[0, 1]$ and therefore, by a result of **[P8]**, is not isomorphic to a separable dual. On the other hand, $X^\perp$ is isometrically isomorphic to the dual of $C(S)/X$. Thus $X^\perp$ is not separable, a contradiction.

(ii) ⇒ (i) **[P4]**. *Case* I: *S metrizable.* If $\mu \in X^\perp$, then, by (ii),

$$\mu = \sum_{s\in S} a_s(\mu)\eta_s$$

where η_s denotes the unit mass at the point s and $a_s(\mu)$ are the complex numbers with $\|\mu\| = \Sigma_{s\in S}|a_s(\mu)|$.

Let $M(X)$ denote the Choquet boundary of X. Then, for every $s \in M(X)$, there is an $x_s \in X$ such that $x_s(s) = 1$ and $|x_s(t)| < 1$ for $t \neq s$ (cf. **[Ph1]**). Hence

$$0 = \lim_n \int_S x_s^n \, d\mu = \lim_n \sum_{t\in S} a_t(\mu) x^n(t) = a_s(\mu).$$

Therefore if $\mu \in X^\perp$, then $|\mu|(M(X)) = 0$. (Here we use the fact that if S is metrizable then $M(X)$ is a G_δ and therefore measurable.) To complete the proof in Case I, it is enough to show that $M(X) = S$. This combined with the previous remark will imply $X^\perp = \{0\}$, i.e., $X = C(S)$.

Suppose to the contrary that there is an s in $S\backslash M(X)$. By the Choquet-Bishop-de Leeuw representation theorem **[Ph1]** there is a $\nu \in [C(S)]^*$ concentrated on $M(X)$ such that

$$\eta_s(x) = x(s) = \int_S x \, d\nu = \int_{M(X)} x \, d\nu \quad \text{for } x \in X.$$

Hence, $\mu = \nu - \eta_s \in X^\perp$, and, by the preceding remark, $\nu(M(X)) = \mu(M(X)) + \eta_s(M(X)) = 0$. On the other hand $\nu(M(X)) = \int_{M(X)} 1 \, d\nu = 1$, a contradiction.

Case II: *S arbitrary compact Hausdorff space.* Then $C(S)/X$ is separable because

$(C(S)/X)^* = X^\perp$ is separable. Hence a standard construction yields the existence of a compact metric space S_1 and a continuous map $h: S \xrightarrow[\text{onto}]{} S_1$ such that if $h^\circ: C(S_1) \longrightarrow C(S)$ is defined by $h^\circ(f) = f \circ h$ and if $X_1 = \{f \in C(S_1): h^\circ(f) \in X\}$ then the map $H: C(S_1)/X_1 \longrightarrow C(S)/X$ defined by $H(\{f + X_1\}) = \{h^\circ(f) + X\}$ is an isometric isomorphism onto $C(S)/X$. Next we check that the operator $(h^\circ)^*: [C(S)]^* \xrightarrow[\text{onto}]{} C(S_1)$ takes purely atomic measures on S onto purely atomic measures on S_1 and restricted to the annihilator of $X^\perp$ coincides with the isometric isomorphism $H^*: X^\perp \xrightarrow[\text{onto}]{} X_1^\perp$. Thus X_1 is an algebra which satisfies the assumption of Case I. Hence $X_1^\perp = \{0\}$ and therefore $X^\perp = \{0\}$. □

REMARK. Chevalier **[Chv]** has extended the argument of Case I. He showed that for every compact Hausdorff space S, $C(S)$ is the only uniform algebra on S with a purely atomic annihilator.

Proposition 5.1 allows us to extend Theorem 4.2 to the case of an arbitrary proper uniform algebra X with a separable annihilator. We replace A by X, H_0^1 by $X^\perp$ (a separable dual!), $L^1(m)/H_0^1$ by $L^1(\lambda)/X^\perp$ where $\lambda \in C(S)_+^*$ is any separable probability measure such that $X^\perp \subset L^1(\lambda)$ and such that whenever $\mu \in L^1(\lambda)$ and μ is singular to $|\nu|$ for every $\nu \in X^\perp$, then $\mu = 0$; H^∞ is replaced by $(X^\perp)^\perp \subset L^\infty(\lambda)$. The natural embedding $i_{m,1}^A$ is replaced by any operator $i_{\mu,1}^X$, where the latter is defined to be T_μ regarded as an operator from X into $X^\perp$ whenever $\mu \in X^\perp$ is chosen so that T_μ is not compact; this is possible, by Proposition 5.1. In particular we have

COROLLARY 5.1. *Let X be a uniform algebra on a compact Hausdorff space S with a norm separable annihilator. Then*

(a) *There is a measure $\mu \in X^\perp$ such that the operator $i_{\mu,1}^X: X \longrightarrow X^\perp$ is absolutely summing, noncompact and non-L^1-factorable.*

(b) *X and all duals of X do not have G-L l.u.st.*

(c) *X is not isomorphic to a quotient of any $C(K)$-space; in particular, X is not complemented in $C(S)$.*

Next we shall show that the properties of the disc algebra exhibited in §4 are shared by a large class of uniform algebras; every uniform algebra with a nontrivial Gleason part belongs to this class. Unfortunately the technical condition which characterizes the class is slightly complicated.

DEFINITION 5.1. Let X be a uniform algebra on S. Then ρ is called a *Hardy measure* for X if there is a triplet (ρ, f_n, F) such that

(5.5) $$\rho \text{ is a probability Borel measure on } S,$$

(5.6) $$f_n \in X, \quad \|f_n\| \leq 1 \quad \text{for } n = 1, 2, \ldots,$$

(5.7) $$F(s) = \lim_n f_n(s) \quad \text{and} \quad |F(s)| = 1 \quad \text{for } \rho\text{-almost all } s,$$

(5.8) $$F\rho \in X^\perp.$$

Our next proposition explains the role of Hardy measures. It is the main technical result of the present section.

PROPOSITION 5.2. *Let X be a uniform algebra on S. Assume that there exists a Hardy measure ρ for X. Then there are operators $T: X \longrightarrow X^{\perp}$ and $Q: X^{\perp} \longrightarrow H_0^1$ such that*

(i) *Q is a contractive projection, that is, $\|Q\| \leqslant 1$ and there is an isometric embedding $J: H_0^1 \longrightarrow X^{\perp}$ such that $QJ = \mathrm{id}_{H_0^1}$.*

(ii) *The natural embedding $i_{m,1}^{A_0}: A_0 \longrightarrow H_0^1$ factors through $(QT)^{**}$, i.e., there are operators $U: A_0 \longrightarrow X^{**}$ and $\pi: (H_0^1)^{**} \longrightarrow H_0^1$ such that $i_{m,1}^{A_0} = \pi(QT)^{**}U$.*

(iii) *T and QT are absolutely summing, noncompact and non-L^1-factorable operators.*

PROOF. Let (ρ, f_n, F) be as in Definition 5.1. Put $T(x) = xF\rho$ for $x \in X$. By (5.8), $T(x) \in X^{\perp}$. Clearly $T: X \longrightarrow X^{\perp}$ is absolutely summing and $\|T\| \leqslant \pi(T) = 1$. The construction of Q is more complicated. Given $f \in C(\partial D)$, $\widetilde{f}$ denotes the continuous function on the unit disc D which extends f and is harmonic for $|z| < 1$. Now fix $n = 1, 2, \ldots$ and assign to every $\nu \in X^{\perp}$ the unique measure $\sigma \in [C(\partial D)]^*$ which represents the linear functional

$$\int_{\partial D} f\, d\sigma = \int_S \widetilde{f}(f_n(s))\nu(ds) \quad \text{for } f \in C(\partial D).$$

Note that, by (5.6), $\widetilde{f} \circ f_n$ is well defined and if $f \in A$ [regarded as a subspace of $C(\partial D)$], then $\widetilde{f}$ is the analytic extension of f onto the interior of the unit disc; hence it admits uniform approximation by polynomials in z. This implies that $\widetilde{f} \circ f_n \in X$; hence $\int_{\partial D} f\, d\sigma = 0$ (because $\nu \in X^{\perp}$). Thus by the M. and F. Riesz theorem, $\sigma = h \cdot m$ for some $h \in H_0^1$. We put $Q_n(\nu) = h$. Clearly $Q_n: X^{\perp} \longrightarrow H_0^1$ is a linear operator with $\|Q_n\| \leqslant 1$, for $n = 1, 2, \ldots$. Let us put

$$Q(\nu) = \operatorname*{Lim}_n Q_n(\nu) \quad \text{for } \nu \in X^{\perp}.$$

Here $\mathrm{Lim}_n h_n$ denotes a fixed Banach limit of a bounded sequence (h_n) with respect to the $\sigma(H_0^1, C(\partial D)/A)$-topology. Since H_0^1 is the dual of $C(\partial D)/A$, the Banach limit exists and $\lim_n h_n$ is a cluster point of the set $\bigcup_{n=1}^{\infty} \{h_n\}$ in the $\sigma(H_0^1, C(\partial D)/A)$-topology. Clearly $Q: X^{\perp} \longrightarrow H_0^1$ is a linear operator with $\|Q\| \leqslant 1$. Moreover

(5.9) if $\nu \in X^{\perp} \cap L^1(\rho)$, then $Q_n(\nu) \longrightarrow Q(\nu)$ in the $\sigma(H_0^1, C(\partial D)/A)$-topology; and if $g = d\nu/d\rho$, then, for every $f \in C(\partial D)$,

$$\int_{\partial D} Q(\nu) \cdot f\, dm = \int_S f(F(s))g(s)\rho(ds).$$

Indeed, given $f \in C(\partial D)$, by (5.7), we have $\lim_n \widetilde{f}(f_n(s)) = \widetilde{f}(F(s)) = f(F(s))$ for ρ-almost all s. Hence, by the Lebesgue theorem,

$$\lim_n \int_{\partial D} Q_n(\nu) \cdot f\, dm = \lim_n \int_S \widetilde{f}(f_n(s))g(s)\rho(ds) = \int_S f(F(s))g(s)\rho(ds).$$

Thus $\int_{\partial D} Q(\nu) f\, dm = \int_S f(F(s))g(s)\rho(ds)$ because a w^*-convergent sequence has a unique w^*-cluster point.

Next we construct an isometric embedding $J: H_0^1 \longrightarrow X^{\perp}$ with $QJ = \mathrm{id}_{H_0^1}$. Assume first that $q \in H_0^1$ is a polynomial, say $q = \sum_{j=1}^k a_j \cdot z^j$. We put $J(q) = q(F) \cdot \rho$. We shall check that

(5.10) $$J(q) \in X^{\perp},$$

(5.11) $$(F^k)_{k=0,\pm 1,\pm 2,\dots} \text{ is an orthonormal sequence in } L^2(\rho),$$

(5.12) $$\|J(q)\|_1 \leqslant \|q\|_1.$$

By (5.8), $F\rho \in X^{\perp}$; if $k > 1$, then, by (5.7), given $x \in X$, $\int_s xF^k\, d\rho = \lim_n \int_S x f_n^{k-1} F\, d\rho = 0$. This proves (5.10). By (5.7), $F(s)^{-1} = \overline{F(s)}$ for ρ-almost all s. Thus, if $k = \pm 1, \pm 2, \dots,$ then $\int F^{-k}\, d\rho = \overline{\int F^k\, d\rho} = 0$. This yields (5.11) because by (5.5) $\int F^k F^{-k}\, d\rho = \int 1\, d\rho = 1$. The verification of (5.12) is a little bit tricky. By (5.11), J extends to an isometric embedding, say $\widetilde{J}$, of H_0^2 into $X^{\perp}$. By the factorization theorem (cf. [**Du**, Chapter 2]), given a polynomial $q \in H_0^1$ there are f_1 and f_2 in H_0^2 with $q = f_1 f_2$ and $\|q\|_1 = \|f_1\|_2 \|f_2\|_2$. Thus

$$\|J(q)\|_1 = \int |q(F)|\, d\rho \leqslant \left(\int |f_1(F)|^2\, d\rho\right)^{1/2} \left(\int |f_2(F)|^2\, d\rho\right)^{1/2}$$
$$= \|\widetilde{J}(f_1)\|_2 \|\widetilde{J}(f_2)\|_2 = \|f_1\|_2 \|f_2\|_2 = \|q\|_1.$$

This proves (5.12). It follows from (5.12) that J extends to an operator (which will be denoted also by J) of norm $\leqslant 1$ from H_0^1 into $X^{\perp}$. To complete the proof of (i) it is enough to show that $QJ(z^k) = z^k$ for $k = 1, 2, \dots$. This will imply that $QJ(h) = h$ for $h \in H_0^1$ and therefore $\|J(h)\| \geqslant \|h\|$, which together with (5.12) yields $\|J(h)\| = \|h\|$. Clearly $J(z^k) = F^k \rho$ for $k = 1, 2, \dots$. Let $f(z) = \Sigma_{j=-r}^{r} a_j z^j$ and let $r > k$. By (5.9) and (5.11) we get

$$\int_{\partial D} Q(F^k \rho) f\, dm = a_{-k} = \int_{\partial D} f(z) z^k m\,(dz).$$

Hence $QJ(z^k) = z^k$ for $k = 1, 2, \dots$. This proves (i).

We identify X with its canonical image in X^{**}. Let $f_* = z^{-1} f$ and $U_n(f) = f_* \circ f_n$ for $f \in A_0$. Since the disc algebra operates on every uniform algebra, $U_n \colon A_0 \longrightarrow X$ is a well-defined linear operator with $\|U_n\| \leqslant 1$, for $n = 1, 2, \dots$. Let us put $U(f) = \operatorname{Lim}_n U_n(f)$ for $f \in A_0$ where "Lim_n" denotes here the Banach limit in the $\sigma(X^{**}, X^*)$-topology. Finally define π to be a contractive projection from $(H_0^1)^{**}$ onto H_0^1. (Since H_0^1 is a dual Banach space, the π in question exists.) Since T is weakly compact, $(QT)^{**}(X^{**}) \subset H_0^1$. Thus $\pi(QT)^{**}(X^{**}) = (QT)^{**}(X^{**}) \in X^{**}$. To complete the proof of (ii) it suffices to show that $(QT)^{**}U(z^k) = z^k$ for $k = 1, 2, \dots$. Fix k. By construction, $U(z^k)$ is a $\sigma(X^{**}, X^*)$-cluster point of the sequence $(U_r(z^k))$. Thus $(QT)^{**}(U(z^k))$ is a $\sigma((H_0^1)^{**}, (H_0^1)^*)$-cluster point of the sequence $(QT^{**}(U_r(z^k))$. Since $(QT)^{**}(X^{**}) \supset H_0^1$, it is in fact a $\sigma(H_1^0, C(\partial D)/A)$-cluster point. We have $(QT)^{**}(U_r(z^k)) = QT(f_r^{k-1}) = Q(f^{r-1} F\rho)$. Hence, by (5.9), for every $f \in C(\partial D)$,

$$\int_{\partial D} Q(f_r^{k-1} F\rho) f\, dm = \int_S f_r^{k-1} F f(F)\, d\rho \qquad \text{for } r = 1, 2, \dots.$$

Now, by (5.7), we infer that the sequence $(Q(f_r^{k-1} F\rho))$ converges in the $\sigma(H_0^1, C(\partial)/A)$-topology to the functional z^k (we have $\lim_r \int_S f_r^{k-1} F f(F)\, d\rho = \int_S F^k f(F)\, d\rho = \int_{\partial D} f(z) z^k\, m(dz)$; the last equality has been verified in the proof of (i)). This proves (ii).

Since T is absolutely summing and $i_{m,1}^{A_0}$ is a noncompact and non-L^1-factorable operator, (iii) follows directly from (ii), Remark R.I in §4 and the fact that a second adjoint of a compact operator is compact. □

COROLLARY 5.2. *If a uniform algebra X admits a Hardy measure, then there is an absolutely summing surjection from X onto l^2.*

PROOF. Let $P: H^2 \longrightarrow H^2$ be a Paley projection such that $P(H^2) \subset H_0^2$. It follows easily from Definition 3.1 that $P(H^2)$ is a closed linear subspace isomorphic to l^2 and P extends to a bounded linear projection, say $\widetilde{P}$, from H^1 onto $P(H^2)$. The desired operator is $\widetilde{P}QT$. It is clearly absolutely summing. By (ii), $(\widetilde{P}QT)^{**}$ is a surjection. Since $\widetilde{P}$ and T are weakly compact and the unit ball of X is dense in the unit ball of X^{**} in the $\sigma(X^{**}, X^*)$-topology, it follows that $\widetilde{P}QT$ maps the unit ball of X onto a set whose norm closure contains a ball of $P(H_0^2)$. Hence $\widetilde{P}QT$ is surjective. □

Combining Proposition 5.2 with an obvious adaptation of the argument used in the proof of Theorem 4.2 we obtain

THEOREM 5.1. *Let X be a uniform algebra on a compact Hausdorff space S. Assume that X admits a Hardy measure. Then*

(j) *X and all the duals of X do not have G-L l.u.st*

(jj) *if X is a quotient of a Banach space Y with G-L l.u.st, then Y contains a complemented isomorph of l^1;*

(jjj) *if $C(S)^*/X^\perp$ is a subspace of a Banach space Z with G-L l.u.st, then Z contains l_n^∞ uniformly;*

(jjjj) *if $C(S)^*/X^\perp$ is isomorphic to a subspace of a Banach lattice Z, then Z contains an isomorph of c_0.* □

COROLLARY 5.3. *If X is a uniform algebra on S which admits a Hardy measure, then X is not complemented in $C(S)$.* □

Our last result gives a sufficient condition in order that a uniform algebra have a Hardy measure, in particular this is the case if a uniform algebra X has a non-one-point Gleason part, i.e., if there are two different homomorphisms of X into complex numbers, say φ and ψ, such that

$$\sup_{\|x\|=1, x\in X} |\varphi(x) - \psi(x)| < 2.$$

PROPOSITION 5.3. *Let X be a uniform algebra on a compact Hausdorff space S. Let us consider the following three properties:*

(1) *X has a Gleason part which contains at least two different points.*

(2) *There exists an ideal $J \subset X$ and a multiplicative linear functional φ on X such that if $a = \|\varphi|J\| = \sup_{f\in J, \|f\|=1} |\varphi(f)|$, then $0 < a < 1$.*

(3) *There is a Hardy measure for X.*

Then (1) $\Rightarrow$ (2) $\Rightarrow$ (3).

PROOF. (1) ⇒ (2). If φ and ψ are in the same Gleason part and $\varphi \neq \psi$, then φ with $J = \ker \psi$ satisfies (2) (cf. [**Gm1**, Chapter VI]).

(2) ⇒ (3). Let μ be a Borel measure on S such that $\int f\, d\mu = \varphi(f)$ for $f \in J$ and $\|\mu\| = a$ (μ exists by the Hahn-Banach theorem). Let $F = d|\mu|/d\mu$ be the Radon-Nikodym derivative of the total variation $|\mu|$ of μ with respect to μ. Let $\rho_1 = |1 - aF|^2 |\mu|$ and $\rho = \rho_1 \|\rho_1\|^{-1}$. Finally let f_n be a sequence in X such that

$$a = \lim_n \varphi(f_n) = \lim_n \int f_n\, d\mu.$$

To complete the proof we shall check that the triplet (ρ, f_n, F) satisfies the conditions (5.5)–(5.8) of Definition 5.1.

Clearly $|F(s)| = 1$ and $1 - a \leqslant |1 - aF(s)| \leqslant 1 + a$ $|\mu|$-almost everywhere. Hence $|\mu|$ is absolutely continuous with respect to ρ and $|F(s)| = 1$ ρ-almost everywhere. Since $\|f_n\| \leqslant 1$ and

$$0 < a = \|\mu\| = \lim_n \int f_n\, d\mu = \lim_n \int \frac{f_n}{F}\, d|\mu|,$$

it follows that $\lim_n f_n(s) = F(s)$ $|\mu|$-almost everywhere and therefore ρ-almost everywhere. Finally given $f \in X$ we have

$$\|\rho_1\| \int fF\, d\rho = \int fF|1 - aF|^2\, d|\mu| = \int fF^2|1 - aF|^2\, d\mu$$

$$= \int fF^2(1 - aF)(1 - a\overline{F})\, d\mu$$

$$= \int [fF^2(1 + a^2) - aF^3 - aF]\, d\mu$$

$$= \lim_n \int f[f_n^2(1 + a^2) - af_n^3 - af_n]\, d\mu$$

$$= \lim_n \varphi(f)[\varphi(f_n)^2(1 + a^2) - a\varphi(f_n)^3 - a\varphi(f_n)] = 0.$$

Thus $F\rho \in X^\perp$. This completes the proof. □

Notes and remarks to §§4 *and* 5. The results of §§4 and 5 are closely related to the following problem due to Glicksberg [**Gl2**].

Problem 5.1. Is $C(S)$ the unique uniform algebra on a compact Hausdorff space S which is complemented in $C(S)$?

Glicksberg (using Rudin's averaging technique [**Ru1**]) proved that the answer is "yes" for S being a homogeneous space of a compact topological group G and for uniform algebras on S which are invariant under the action of G on $C(S)$: $(T_g f)(s) = f(gs)$ for $f \in C(S)$, $g \in G$, $s \in S$. For details and generalization to locally compact groups cf. [**Gl2**] and [**R1**]. It was shown in [**P4**] that the answer is "yes" for an arbitrary compact Hausdorff space S and for uniform algebras on S with norm separable annihilators (cf. our Corollary 5.1). For uniform algebras with nontrivial Gleason parts the same fact (cf. our Corollary 5.3) was proved by Kisliakov [**Kis1**] and with some additional assumptions by Etcheberry [**Etch**]. While Glicksberg's problem is still open, it seems to be more natural to ask

Problem 5.2. *Conjecture.* Let X be a proper uniform algebra on a compact Hausdorff space S (i.e., $C(S)\backslash X$ is nonempty). Then

(a) X as a Banach space is not isomorphic to any $C(K)$-space

(b) X is not isomorphic to any complemented subspace of a $C(K)$-space

(c) X is not isomorphic to any quotient space of a $C(K)$-space

(d) X does not have G-L l.u.st.

It was observed in **[P5]** that if S is metrizable and X is a uniform algebra on S which is complemented in a $C(K)$-space then X is isomorphic as a Banach space to $C(S)$. Varopoulos **[Va]** proved recently that if a semisimple algebra X is isomorphic as a Banach space to a $C(K)$-space, then X is an operator algebra. By a result of Bishop (cf. **[B2]**, **[Gm1**, pp. 60–62**]**), it follows that every uniform algebra has a quotient antisymmetric algebra; thus in (c) one may assume without loss of generality that X is antisymmetric.

It is not clear whether the criterion of Proposition 5.2 can be applied for uniform algebras with norm separable annihilators. This suggests the following

Problem 5.3. Let X be a uniform algebra with a norm separable annihilator. Does X admit a Hardy measure? Does X have a non-one-point Gleason part?

Another approach to attacking Problem 5.2 is to eliminate the assumption of the norm separability of the annihilator in Proposition 5.1 (cf. the Remark after the proof of Proposition 5.1). This suggests

Problem 5.4. Let X be a proper uniform algebra on S. Does there exist a $\mu \in X^{\perp}$ such that the operator T_μ: $X \longrightarrow L^1(|\mu|)$ is not compact where $T_\mu(x) = x d\mu/d|\mu|$ for $x \in X$?

Theorem 4.2(i) was proved in **[P7]** while (ii) and (iii) were observed in **[P6]**, (iv) was first proved using a different technique of **[Pis1]** by Pisier, (v) is due to W. B. Johnson (oral communication). The equivalence (iii) $\Longleftrightarrow$ (i) of Proposition 5.1 is due to Wojtaszczyk **[W1]**, while (i) $\Longleftrightarrow$ (ii) is taken from **[P4]**. The concept of a Hardy measure is implicitly contained in Kisliakov's work **[Kis1]**. Proposition 5.2 in the present form was stated and proved by Wojtaszczyk. A slightly weaker fact was established in **[P7]**. Proposition 5.3 goes back to Bishop and Wermer (cf. **[Gm1**, Chapter VI, proof of Theorem 7.1**]**).

ADDED IN PROOF. A similar result to Proposition 5.2 has been obtained by Kisliakov **[Kis3]**.

6. Uniformly Peaking Families of Functions in A and H^∞. The Havin Lemma.

This section has a preparatory character for the next one. The main result is a very technical Proposition 6.1 due to Havin **[Hv1]**. This is a generalization of a quantitative character of the fact that every closed subset of the unit circle of Lebesgue measure zero is a peak set for the disc algebra (cf. §1, Step I in the proof of the F. and M. Riesz theorem). Roughly speaking it is shown that if a closed subset e of the unit circle has a small Lebesgue measure then there is a function in the disc algebra of norm one whose values are close to zero except for a set of small measure and are close to one at each point of e. Moreover all the estimations on the measure of the exceptional set and how the values of the function are close either to zero or to one depend on the measure of e only. A similar construction leads to a result of Amar and Lederer **[A-L]** on peak sets for H^∞ (Proposition 6.2) and to a characterization of exposed points of H^∞ (Theorem 6.1).

Recall that m denotes the normalized Lebesgue measure on the unit circle $\partial D = \{z \in \mathbf{C}\colon |z| = 1\}$. Given a measurable function $u\colon \partial D \longrightarrow \mathbf{C}$, we denote (if it exists) by $H(u)$ the Hilbert transform of u (cf. §0.II) and by $\tilde{u}$ the harmonic extension of u onto D, i.e.,

$$\tilde{u}(re^{i\theta}) = (2\pi)^{-1}\int_0^{2\pi} u(e^{i\theta})\,\frac{1-r^2}{1-2\,r\cos\theta + r^2}\,d\theta \quad \text{for } re^{i\theta} \in D\backslash\partial D.$$

PROPOSITION 6.1 (THE HAVIN LEMMA). *There exists a positive function $\delta \longrightarrow \epsilon(\delta)$ for $0 < \delta < 1$ with $\lim_{\delta=0}\epsilon(\delta) = 0$ such that given a measurable subset e of ∂D with $0 < m(e) < \delta$ there exist f_e and g_e in H^∞ with the following properties*:

(6.1) $$|f_e(z)| + |g_e(z)| = 1 \quad \textit{for } z \in \partial D,$$

(6.2) $$\sup_{z\in e} |f_e(z) - 1| = O(\delta^{1/2}), \qquad \int_{\partial D} |f_e(z)|\,dz = O(\delta^{1/6}),$$

(6.3) $$\sup_{z\in e} |g_e(z)| = O(\delta^{1/2}), \qquad \int_{\partial D} |g_e(z) - 1|\,dz < \epsilon(\delta),$$

(6.4) *if e is closed, then f_e and g_e are in A.*

PROOF. *Step* I. *Construction.* Fix $e \subset \partial D$ measurable with $m(e) < \delta \leqslant 1$ and put $B = \delta^{-1/2}$, $b = \delta^{1/2}$. If e is not closed we define $w_e = w$ by

$$w(z) = \begin{cases} -B & \text{if } z \in e, \\ -b & \text{if } z \notin e. \end{cases}$$

If e is closed we first pick finite families of closed intervals $(\Delta_j')_{1 \leqslant j \leqslant N}$ and $(\Delta_j'')_{1 \leqslant j \leqslant N}$ such that if $e' = \bigcup_{j=1}^N \Delta_j'$ and $e'' = \bigcup_{j=1}^N \Delta_j''$ then

$$(6.5) \qquad \begin{gathered} e \subset e', \quad m(e'') < \delta, \quad \Delta_j' \subset \operatorname{int} \Delta_j'' \quad \text{and} \\ \Delta_k' \cap \Delta_j' = \Delta_k'' \cap \Delta_j'' = \varnothing \quad \text{for } k \neq j \ (k, j = 1, 2, \ldots, N), \end{gathered}$$

and we define w to be a C^∞-function on ∂D such that $-B \leqslant w(z) \leqslant -b$ for all $z \in \partial D$ and

$$w(z) = \begin{cases} -B & \text{for } z \in e', \\ -b & \text{for } z \notin e''. \end{cases}$$

Next we put

$$v = H(w); \qquad h = w + iv; \qquad f_e = \exp h^{-1}.$$

We shall check

$$(6.6) \qquad \begin{gathered} \widetilde{h}(z) \neq 0 \quad \text{for } z \in D, \qquad |f_e(z)| \leqslant 1 \quad \text{for } z \in \partial D, \\ \int_{\partial D} \log(1 - |f_e|)\, dm > -\infty. \end{gathered}$$

Then we define g_e to be the outer function with $|g_e(z)| = 1 - |f_e(z)|$ for $z \in \partial D$ almost everywhere. Precisely we put

$$(6.7) \qquad \widetilde{g}_e(z) = \exp \frac{1}{2\pi} \int_0^{2\pi} \frac{e^{it} + z}{e^{it} - z} \log(1 - |f_e(e^{it})|\, dt \quad \text{for } |z| < 1,$$

$$g_e(z) = \lim_{\rho = 1} \widetilde{g}_e(\rho z) \quad \text{for } z \in \partial D \text{ almost everywhere.}$$

Step II. *Verification of* (6.6). By the maximum principle for harmonic functions,

$$-B \leqslant \operatorname{Re} \widetilde{h}(z) = \widetilde{w}(z) \leqslant -b \quad \text{for } z \in D.$$

Hence $\widetilde{h}(z) \neq 0$ for $z \in D$ and

$$\operatorname{Re} \widetilde{h}^{-1} = \frac{\widetilde{w}}{\widetilde{w}^2 + \widetilde{v}^2} \leqslant -b < 0.$$

Since $w \in L^\infty$, $v = H(w)$ exists and $v \in L^p$ for all p (cf. §0.II). Thus $\widetilde{h}$ and $\widetilde{h}^{-1}$ exist and are analytic for $|z| < 1$. Furthermore $|e^{\widetilde{h}^{-1}}| \leqslant e^{\operatorname{Re} \widetilde{h}^{-1}} \leqslant 1$; hence $f_e \in H^\infty$ and $|f_e(z)| \leqslant 1$ for $z \in \partial D$ almost everywhere. It remains to show that $\int_{\partial D} (1 - |f_e|)\, dm > -\infty$; this will imply that g_e defined by (6.7) does exist and belongs to H^∞ (cf. [**Du**, Chapter 2]).

Let $Z_0 = \{z \in \partial D\colon v(z)^2 \leqslant 2\}$ and $Z_k = \{z \in \partial D\colon 2^k \leqslant v^2(z) < 2^{k+1}\}$ for $k = 1, 2, \ldots$. Observe that there is a $C > 0$ such that $m(Z_k) < C 2^{-k}$ for $k = 0, 1, \ldots$ because $v \in L^2$. If $z \in Z_k$, then

$$|\exp h^{-1}(z)| = \exp \frac{w(z)}{w^2(z) + v^2(z)} \geqslant \exp \frac{-b}{B^2 + 2^{k+1}} \geqslant 1 + \frac{b}{B^2 + 2^{k+1}}.$$

Thus

$$\int \log(1 - |f|)\, dm = \sum_{k=0}^{\infty} \int_{Z_k} \log(1 - |\exp h^{-1}(z)|)\, dz$$

$$\geqslant \sum_{k=0}^{\infty} \log\left(\frac{b}{B^2 + 2^{k+1}}\right) \cdot m(Z_k)$$

$$\geqslant C \sum_{k=0}^{\infty} 2^{-k} \log \frac{b}{B^2 + 2^{k+1}} > -\infty.$$

Step III. *Verification of* (6.4). If e is closed, then w and therefore v are C^∞-functions (cf. §0.II). Hence h, h^{-1} and f_e are in A. Moreover $0 < |f_e(e^{it})| < 1$ is a 2π-periodic C^∞-function, hence $g_e \in A$ (cf. [**Du**, Chapter 5]).

Step IV. *Verification of* (6.1) *and* (6.2). The formula (6.7) yields that $|g_e(z)| = 1 - |f_e(z)|$ for $z \in \partial D$ almost everywhere (cf. [**Du**, Chapter 2]). This yields (6.1). If $z \in e$, then

$$\frac{1}{|h(z)|} = \frac{1}{\sqrt{w^2(z) + v^2(z)}} \leqslant \frac{1}{|w(z)|} = B^{-1}.$$

Hence remembering that $B > 1$, we get

$$|1 - f_e(z)| = |1 - \exp h^{-1}(z)| \leqslant \exp|h^{-1}(z)| - 1$$
$$\leqslant |h^{-1}(z)| \exp|h^{-1}(z)| \leqslant B^{-1}\exp(1) = O(\delta^{1/2}).$$

The proof of the second inequality of (6.2) is slightly harder. To treat simultaneously the case where e is a closed set and e is not closed we put $e^* = e''$ if e is closed, and $e^* = e$ otherwise. Let $X = \{z \in \partial D\colon |v(t)| \geqslant \eta\}$ for some $\eta > 0$. Since the Hilbert transform is of the weak type (1.1) there is an absolute constant C_1 such that

$$m(X) \leqslant C_1 \eta^{-1} \int_{\partial D} |w|\, dm \leqslant C_1 B\delta\eta^{-1} + C_1 b\eta^{-1},$$

because $\int_{\partial D} |w|\, dm = \int_{\partial D \setminus e^*} |w|\, dm + \int_{e^*} |w|\, dm \leqslant B\delta + b$. Thus, taking into account that if $z \notin X \cup e^*$ then

$$|f_e(z)| = \exp \operatorname{Re} h^{-1}(z) = \exp \frac{w(z)}{w^2(z) + v^2(z)}$$
$$\leqslant \exp \frac{-b}{b^2 + \eta^2} \leqslant b + \eta^2 b^{-1},$$

we get

$$\int_{\partial D} |f_e|\, dm = \int_{e^*} |f_e|\, dm + \int_{X \setminus e^*} |f_e|\, dm + \int_{\partial D \setminus X \cup e^*} |f_e|\, dm$$
$$\leqslant m(e^*) + C_1(B\eta^{-1}\delta + b\eta^{-1}) + b + \eta^2 b^{-1}$$
$$\leqslant O(\delta^{1/6}),$$

provided we put $\eta = \delta^{1/3}$. This completes Step IV.

Step V. *Verification of* (6.3). Let $u_e = |\log(1 - |f_e|)|$. Then $g_e = \exp(-u_e - iH(u_e))$. The crucial point is to show that there is a constant M such that

(6.8) $$\int_{\partial D} u_e^2\, dm < M \quad \text{for all measurable } e < \partial D.$$

Having done this we complete the proof as follows: By (6.1) and (6.2), $u_e \longrightarrow 0$ in measure whenever $m(e) \longrightarrow 0$. Thus, by (6.10), $\int_{\partial D} |u_e|\, dm \longrightarrow 0$ whenever $m(e) \longrightarrow 0$. Since the Hilbert transform is of weak type (1.1), the last relation yields

$$m\left\{z \in \partial D\colon |H(u_e)(z)| \geqslant \left(\int_{\partial D} |u_e|\, dm\right)^{1/2}\right\} \leqslant C_1 \left(\int_{\partial D} |u_e|\, dm\right)^{1/2}$$

where C_1 is an absolute constant (cf. §0.II). Thus if $m(e) \longrightarrow 0$ then $H(u_e) \longrightarrow 0$ in measure and $g_e \longrightarrow 1$ in measure. Since $\|g_e\|_\infty \leqslant 1$ for all measurable $e < \partial D$, we conclude that $\int |g_e - 1|\, dm \longrightarrow 0$ whenever $m(e) \longrightarrow 0$ which proves the second inequality of (6.3). The first formula of (6.3) follows directly from (6.1) and the first formula of (6.2).

To establish (6.8) put

$$y = \left|\operatorname{Re} \frac{1}{h}\right| = \frac{|w|}{w^2 + v^2}.$$

Clearly $0 < y < \infty$ and

$$\begin{aligned} u_e &= \log \frac{1}{1 - \exp(-y)} \leqslant 6\left(\frac{1}{1 - \exp(-y)}\right)^{1/6} \\ &= 6\left(1 + \frac{1}{\exp(y) - 1}\right)^{1/6} \leqslant 6(1 + y^{-1})^{1/6}. \end{aligned}$$

Therefore $u_e^2 \leqslant 36\,(1 + y^{-1})^{1/3} \leqslant 36\,(1 + y^{-1/3})$. Finally

$$\int y^{-1/3}\, dm = \int \left(\frac{v^2}{|w|} + |w|\right)^{1/3} dm \leqslant \int \frac{v^{2/3}}{|w|^{1/3}}\, dm + \int |w|^{1/3}\, dm.$$

We estimate each term separately. By Kolmogorov's theorem (cf. §0.II), there is an absolute constant $C_{2/3}$ such that $\int |v|^{2/3}\, dm \leqslant C_{2/3}(\int |w|\, dm)^{2/3}$. Hence

$$\begin{aligned} \int \frac{v^{2/3}}{w^{1/3}}\, dm &\leqslant b^{-1/3} C_{2/3} \left(\int |w|\, dm\right)^{2/3} \\ &\leqslant C_{2/3} b^{-1/3} (B\delta + b)^{2/3} \\ &\leqslant C_{2/3} \delta^{-1/6} (2\delta^{1/2})^{2/3} \leqslant 2^{2/3} C_{2/3}, \end{aligned}$$

while $\int |w|^{1/3}\, dm \leqslant B^{1/3}\delta + b^{1/3} \leqslant 2$. □

REMARK. The estimation for the order of the function $\delta \longrightarrow \epsilon(\delta)$ which follows from the argument in Step V seems to be unsatisfactory. We get only $\epsilon(\delta) \leqslant C|\log|1 - e\delta^a|\,|^{2a}$ for $a < 1/2$.

The next result, due to Amar and Lederer [A-L], uses a similar technique to that of Havin's lemma. It is, however, formulated in a different language, which we now recall.

Let Δ be the maximal ideal space of L^∞ and let $f \longrightarrow \hat{f}$ be the algebraic isometric isomorphism from L^∞ onto $C(\Delta)$. Recall that Δ is a compact Hausdorff space. If χ_e is the characteristic function of a measurable subset e of ∂D, then $\hat{e} = \{t \in \Delta\colon \hat{\chi}_e(t) = 1\}$ is a clopen set; all the sets $\hat{e}$ form a basis for the topology of Δ. By $\hat{m}$ we denote the unique regular Borel measure on Δ which corresponds via the Riesz representation theorem to the functional $\hat{f} \longrightarrow \int f\,dm$, i.e., $\hat{m}$ is defined by

$$\int \hat{f}\,d\hat{m} = \int f\,dm \quad \text{for } f \in L^\infty.$$

H^∞ can be regarded as the uniform algebra on Δ, in fact Δ is the Shilov boundary for H^∞ (cf. [**H**, Chapter 10]). Now we are ready to state

PROPOSITION 6.2. *Let F be a closed G_δ in Δ with $\hat{m}(F) = 0$. Then F is contained in a nontrivial peak set for H^∞, i.e., there exists a nonconstant $f \in H^\infty$ such that $\|f\| = 1$, $F \subset \hat{f}^{-1}(1) = \{t \in \Delta: \hat{f}(t) = 1\}$ and $\hat{f}^{-1}(1) = \{t \in \Delta: |\hat{f}(t)| = 1\}$.*

PROOF. The assumption on F yields the existence of a decreasing sequence (e_n) of measurable subsets of ∂D such that $F = \bigcap_{n=1}^\infty \hat{e}_n$ and $m(e_n) = \hat{m}(\hat{e}_n) \longrightarrow 0$ as $n \longrightarrow \infty$. Passing to a subsequence, if necessary, we may assume without loss of generality that $\hat{e}_1 = \Delta$, $\hat{m}(\hat{e}_n) < n^{-3}$ for $n = 1, 2, \ldots$. Now we put

$$w = \sum_{n=1}^\infty -n\,\chi_{e_n \setminus e_{n+1}}, \qquad v = H(w),$$

$$h = w + iv, \qquad f = \exp(1/h).$$

Since $w \in L^p$ for $p < 2$, $h \in H^p$. Thus h is a boundary value of an analytic function, say $\widetilde{h}$ in the unit disc. Furthermore, $\operatorname{Re} \widetilde{h} \leq -1$, because $\operatorname{Re} h \leq -1$. Thus $\widetilde{h} \neq 0$ and f is a nonconstant H^∞ function with $\|f\|_\infty \leq 1$. Furthermore if $t \in F$ then $\hat{f}(t) = 1$ because for $z \in e_n$, $|f(t) - 1| = |\exp(1/h) - 1| \leq |\exp(1/|h|) - 1| \leq e^{1/n} - 1$. Hence

$$\hat{f}^{-1}(1) = \bigcap_{n=1}^\infty \{t \in \Delta: |\hat{f}(t) - 1| \leq e^{1/n} - 1\} \supset \bigcap_{n=1}^\infty \hat{e}_n = F.$$

Finally replacing if necessary f by $\frac{1}{2}(f + 1)$, we obtain the last assertion of Proposition 6.2. □

It is interesting to compare Proposition 6.2 with the next one.

PROPOSITION 6.3. *There is a closed nonempty G_δ-set $F \subset \Delta$ with $\hat{m}(F) = 0$ which is not a peak set for H^∞.*

PROOF. Suppose that the assertion of Proposition 6.3 is false. Then given a closed G_δ-set $F \subset \Delta$ with $\hat{m}(F) = 0$ there exists an $f_F \in H^\infty$ such that $\hat{f}_F$ peaks exactly on F. Hence

(6.9) $$\text{if } \mu \in (H^\infty)^\perp, \text{ then } \mu \ll \hat{m}.$$

Indeed, by regularity of μ, it is enough to observe that $\mu(F) = 0$ for every closed G_δ-set $F \subset \Delta$ with $\hat{m}(F) = 0$. The last follows from

$$\mu(F) = \lim_n \int \hat{f}_F^n\,d\mu = 0$$

(because $\hat{f}_F^n$ converges pointwise to the characteristic function of F). The map $f \longrightarrow \hat{f}$ for $f \in L^\infty$ induces an isometric isomorphism from $L^1(m)$ onto $L^1(\hat{m})$ which carries H_0^1 onto

$L^1(\hat{m}) \cap (H^\infty)^\perp$. Now (6.9) implies that the annihilator $(H^\infty)^\perp$ coincides with $L^1(\hat{m}) \cap (H^\infty)^\perp$; hence $(H^\infty)^\perp$, being isometrically isomorphic to H_0^1, is norm separable. Since $(L^\infty/H^\infty)^*$ is isometrically isomorphic to $(H^\infty)^\perp$, we infer that (L^∞/H^∞) is separable. This leads to a contradiction because $(H_0^1)^*$ is isometrically isomorphic to L^∞/H^∞ and H_0^1 contains a subspace isomorphic to l^1. □

Problem 6.1. Characterize the subsets of the Shilov boundary of H^∞ which are peak sets for H^∞.

We end this section with an application of Proposition 6.2 of a geometric character. We shall characterize the exposed points of the unit ball of H^∞.

Recall that if X is a Banach space then an $x \in X$ with $\|x\| = 1$ is an *exposed point* of the unit ball $B_X = \{y \in X: \|y\| \leqslant 1\}$ if there is an $x^* \in X^*$ with $\|x^*\| = 1$ which *strictly supports* B_X at x, i.e., $x^*(x) = 1$ and $\operatorname{Re} x^*(y) < 1$ for every $y \in B_X \setminus \{x\}$.

THEOREM 6.1. *A function $f \in H^\infty$ with $\|f\| = 1$ is an exposed point of the unit ball B_{H^∞} iff*

(6.10) $$m(E) > 0 \quad \textit{where } E = \{z \in \partial D : |f(z)| = 1\}.$$

PROOF. Let f be an exposed point for B_{H^∞} and let $x^* \in (H^\infty)^*$ be a linear functional which strictly supports B_{H^∞} at f. Let $\Phi^* \in (L^\infty)^*$ be a norm preserving extension of x^*. Given measurable $e \subset \partial D$ we denote by Φ_e^* the linear functional on L^∞ defined by $\Phi_e^*(g) = \Phi^*(\chi_e g)$ for $g \in L^\infty$. Clearly $\Phi^* = \Phi_e^* + \Phi_{\partial D\setminus e}^*$ and $\|\Phi^*\| = \|\Phi_e^*\| + \|\Phi_{\partial D\setminus e}^*\|$. Furthermore if e satisfies the condition $|f(z)| \leqslant a < 1$ for $z \in \partial D\setminus e$ almost everywhere, then $\Phi_e^* = \Phi^*$. Indeed, otherwise $\|\Phi_{\partial D\setminus e}^*\| > 0$ and

$$\|\Phi^*\| = \Phi^*(f) \leqslant |\Phi_e^*(f)| + |\Phi_{\partial D\setminus e}^*(f)|$$

$$\leqslant \|\Phi_e^*\| + a\|\Phi_{\partial D\setminus e}^*\| < \|\Phi^*\|,$$

a contradiction.

Now let $e_n = \{z \in \partial D: 1 - |f(z)| < n^{-1}\}$. By the previous remark $\Phi_{e_n}^* = \Phi^*$ for $n = 1, 2, \ldots$. Clearly $e_1 \supset e_2 \supset \cdots$. Let $e_\infty = \bigcap e_n$. To show that f satisfies (6.10) it is enough to establish that $m(e_\infty) > 0$. Suppose that $m(e_\infty) = 0$; then $F = \bigcap_{n=1}^\infty \hat{e}_n$ is a closed G_δ-subset of Δ with $\hat{m}(F) = 0$. Thus, by Proposition 6.2, there is a nonconstant $f_F \in H^\infty$ with $\|f_F\| = 1$ such that $\hat{f}_F^{-1}(1) \supset F$. For $n = 1, 2, \ldots$, we have

$$|\Phi^*(f) - \Phi^*(f_F f)| = \Phi_{e_n}^*((1 - f_F)f) \leqslant \|f\| \sup_{z\in e_n} |1 - f_F(z)|.$$

Taking into account that $\hat{e}_1 \supset \hat{e}_2 \supset \cdots \supset F = \bigcap_{n=1}^\infty \hat{e}_n$ and $\hat{f}_F^{-1}(1) \supset F$, we infer that $\lim_n \sup_{z\in e_n} |1 - f_F(z)| = 0$. Hence $\Phi^*(f) = \Phi^*(f \cdot f_F)$ which contradicts the definition of the exposed point, because f_F being nonconstant implies that $f \cdot f_F \in B_{H^\infty} \setminus \{f\}$. Hence $m(e_\infty) > 0$.

Now assume that an $f \in H^\infty$ with $\|f\| = 1$ satisfies (6.10). Let $\Phi^* \in (L^\infty)^*$ be defined by

$$\Phi^*(g) = m(E)^{-1} \int_E g(z) \operatorname{sign} f(z) m(dz) \quad \text{for } g \in L^\infty.$$

Let x^* be the restriction of Φ^* onto H^∞. Clearly $\|x^*\| = |x^*(f)| = \|\Phi^*\| = 1$. Hence $\operatorname{Re} x^*(g) \leqslant 1$ for $g \in B_{H^\infty}$. Finally if, for some $g \in B_{H^\infty}$, $\Phi^*(g) = x^*(g) = 1$, then $|g(z)| = 1$ and $g(z)\operatorname{sign} f(z) = 1$ for $z \in E$ almost everywhere. Hence $g(z) = f(z)$ for $z \in E$ almost everywhere. Since $m(E) > 0$, we infer that $f = g$. □

REMARK. Note that if f is an exposed point for B_{H^∞} then a linear functional which strictly supports B_{H^∞} at f can be chosen from L^1/H^1_0, a predual of H^∞.

Notes and Remarks to §6. The proof of Havin's lemma is taken with some modification from **[Hv1]**. The condition (6.4) of the lemma was observed by Kisliakov **[Kis2]** and independently by Pelczynski (unpublished). Wojtaszczyk **[W2]** observed recently that Havin's lemma can be generalized to uniform algebras with unique representing measures for linear multiplicative functionals. The proof of Proposition 6.2 is taken from **[A-L]**. The construction is classical (cf. **[Z**, Vol. I, p. 105**]**). Proposition 6.3 shows that the "Remarque" in **[A-L]** is false. Theorem 6.1 is due to Amar and Lederer **[A-L]** and Fisher **[Fi]**. An analogous result for the disc algebra is due to Phelps **[Ph2]**.

Let us recall that an $f \in B_{H^\infty}$ is an extreme point of B_{H^∞} iff $\int_{\partial D} \log(1 - |f|)\, dm = -\infty$ (cf. **[H**, p. 138**]**). Finally observe that no point of B_{H^∞} is *strongly exposed*, i.e., there is no $f \in B_{H^\infty}$ and $x^* \in (H^\infty)^*$ with $x^*(f) = \|f\| = \|x^*\| = 1$ and such that for every sequence (f_n) in B_{H^∞} if $x^*(f_n) \longrightarrow x^*(f)$ then $\|f_n - f\| \longrightarrow 0$. Indeed, regard H^∞ as a subspace of $C(\Delta)$ and note that for every $\mu \in C(\Delta)^*$ there exists a peak set, say F, for H^∞ with $\mu(F) = 0$. Suppose $f_F \in H^\infty$ peaks exactly at F, i.e.,

$$\{s \in \Delta: \hat{f}_F(s) = 1\} = \{s \in \Delta: |\hat{f}_F(s)| = 1\} = F.$$

Given $f \in B_{H^\infty}$ put $f_n = f(1 - f_F^n)/\|1 - f_F^n\|$ for $n = 1, 2, \ldots$. Then $\lim_n \|f_n - f\| = \|f\|$ for $n = 1, 2, \ldots$ while $\lim_n \int \hat{f}_n\, d\mu = \int \hat{f}\, d\mu$.

Various interesting properties of H^∞ and L^1/H^1_0 related to the material of this section are discussed in the expository paper by Havin **[Hv2]**.

ADDED IN PROOF. Delbaen **[De3]** recently generalized Havin's Lemma for a representing measure of a linear multiplicative functional on a uniform algebra such that the representing measures of the functional form a weakly compact set.

7. Characterizations of Weakly Compact Sets in L^1/H_0^1 and in A^*

In this and in the next section we shall show that, despite the results of §4, the spaces A, A^* and L^1/H_0^1 share various properties of $C(\partial D)$, $C(\partial D)^*$, and L^1. In particular a specific characterization of weakly compact sets in L^1-spaces **[Gr2]**, **[P9]** as those sets on which weakly unconditional series of functionals converge uniformly can be carried out to L^1/H_0^1 and A^*. Another peculiar property of a weakly compact set in L^1/H_0^1 says that the set is an image under the quotient map of a weakly compact set in L^1. The main result of this section, Theorem 7.1, was proved recently by Delbaen **[D1]** and independently by Kisliakov **[Kis2]**.

Recall that a sequence (x_n) of elements of a Banach space X is w.u.s. (*weakly unconditionally summable*) if $\Sigma_{n=1}^{\infty}|x^*(x_n)| < \infty$ for every $x^* \in X^*$. Clearly if (x_n) is w.u.s. then there is a $K > 0$ such that $\Sigma|x^*(x_n)| \leqslant K\|x^*\|$ for $x^* \in X^*$. A sequence (y_n) in X is *weakly Cauchy* if $\lim_n x^*(y_n)$ exists for every $x^* \in X^*$. The following classical result goes back to Orlicz.

LEMMA 7.1. *Let W be a subset of X such that every sequence of elements of W contains a weak Cauchy subsequence. Then*

(7.1) $$\lim_n \sup_{x\in W} |x_n^*(x)| = 0 \quad \textit{for every w.u.s. sequence } (x_n^*) \textit{ in } X^*.$$

PROOF. Define $T\colon X \longrightarrow l^1$ by $Tx = (x_n^*(x))$. Then T is a bounded linear operator which carries W into a totally bounded subset of l^1, because in l^1 every weakly Cauchy sequence is norm convergent. Hence $\lim_N \Sigma_{n=N}^{\infty}|x_n^*(x)| = 0$ uniformly for $x \in W$. This yields (7.1). □

In view of formulas (1.1) and (1.2) of §1, in this and in the next section we identify L^1/H_0^1 with a subspace of A^* and A with a subspace of $(L^1/H_0^1)^*$. The duality is given by the bilinear form (B_E denotes the unit ball of a Banach space E)

$$(\omega, x) \longrightarrow \int_{\partial D} xw\, dm \quad \text{for } x \in A \text{ and } \omega = \{w + H_0^1\} \in L^1/H_0^1.$$

Moreover, $\sup_{x\in B_A}\int_{\partial D} xw\, dm = \|\omega\|_{L^1/H_0^1}$ and $\sup_{\omega\in B_{L^1/H_0^1}}\int_{\partial D} xw\, dm = \|x\|_A$.

Let Y be a closed linear subspace of a Banach space X. A map (in general nonlinear) $\tau\colon X/Y \longrightarrow X$ is called the *nearest point cross-section* if

$$\tau(\omega) \in \omega \quad \text{and} \quad \|\omega\|_{X/Y} = \inf_{y\in Y} \|y + \tau(\omega)\|_X = \|\tau(\omega)\|_X \quad \text{for } \omega \in X/Y.$$

The nearest point cross-section need not exist. It does in the case where Y is H_0^1 and X is either L^1 or X is $C(\partial D)^*$, because by the F. and M. Riesz theorem, H_0^1 is a closed subspace of $C(\partial D)^*$ in the $\sigma(C(\partial D)^*, C(\partial D))$-topology. In fact in these cases it is unique (cf. **[Kh1]**, **[Hv2]**).

Now we are ready to state

THEOREM 7.1. *Let W be a subset of L^1/H_0^1. Then the following conditions are equivalent:*

(1) *There is a weakly compact subset V of L^1 such that $q(V) \supset W$ where $q\colon L^1 \to L^1/H_0^1$ is the quotient map. More precisely, for V one can take the weak closure of the image of W under the nearest point cross-section.*

(2) *The weak closure of W in L^1/H_0^1 is weakly compact.*

(3) *Every sequence of elements of W contains a weak Cauchy subsequence.*

(4) $\lim_n(\sup_{\{w+H_0^1\}\in W}|\int_{\partial D}\varphi_n w\,dm|) = 0$ *for every w.u.s. sequence (φ_n) in A.*

PROOF. Clearly for arbitrary Banach space (1) ⇒ (2) ⇒ (3) ⇒ (4). (Use Lemma 7.1 and the "easier" implication of the Eberlein-Šmulian theorem.) To prove that (4) ⇒ (1), we observe first that without loss of generality one may assume

(7.2) $\qquad W$ is an infinite countable set,

(7.3) $\qquad$ every $\{w + H_0^1\} \in W$ regarded as a linear functional on A attains its norm on the unit ball of A.

For (7.2) note that L^1/H_0^1 is separable and observe that if a subset of L^1/H_0^1 satisfies (1) so does its closure.

For (7.3) note: 1° By the open mapping principle, if $(\{w_n' + H_0^1\})$ and $(\{w_n'' + H_0^1\})$ are enumeration of countable sets W' and W'' respectively and if $\|\{w_n' + H_0^1\} - \{w_n'' + H_0^1\}\|_{L^1/H_0^1} < 2^{-n}$ for $n = 1, 2, \ldots$, then W' satisfies (1) iff W'' does. 2° By the Bishop-Phelps theorem **[B-Ph]**, the subset A_{BP}^* of A^* consisting of the linear functionals which attain their norms on the unit ball of A is dense in A^*. Since A^* is the l^1-sum of L^1/H_0^1 and V_{sing} (cf. §1), the natural projection of A^* onto L^1/H_0^1 which annihilates V_{sing} maps A_{BP}^* into itself. Hence $A_{BP}^* \cap L^1/H_0^1$ is dense in L^1/H_0^1.

Now let $(\{w_n + H_0^1\})$ be an enumeration of W. It follows from (7.3) that there is a $y_n \in A$ so that

$$\int_{\partial D} w_n y_n\,dm = \inf_{h\in H_0^1}\int_{\partial D}|w_n + h|\,dm \quad\text{and}\quad \|y_n\| = 1 \text{ for } n = 1, 2, \ldots. \tag{7.4}$$

By the F. and M. Riesz theorem, H_0^1 is a closed subspace of $C(\partial D)^*$ in the $\sigma(C(\partial D)^*, C(\partial D))$-topology; hence there are $h_n \in H_0^1$ such that

$$\inf_{h\in H_0^1}\int_{\partial D}|w_n + h|\,dm = \int_{\partial D}|w_n + h_n|\,dm \quad \text{for } n = 1, 2, \ldots. \tag{7.5}$$

Let $v_n = w_n + h_n$ for $n = 1, 2, \ldots$ and let $V = \bigcup_{n=1}^\infty \{v_n\}$. Our goal is to show

(7.6) the weak closure of V in L^1 is weakly compact.

Since $q(V) = W$ and $\tau(\{w_n + H_0^1\}) = v_n$ for $n = 1, 2, \ldots$, this will complete the proof.

We left to the reader a simple checking that if W satisfies (4) then it is norm bounded. This implies that $\sup_n \|v_n\| = M < \infty$. To apply the well-known criterion of weak compactness in L^1 it remains to show that the v_n's are uniformly integrable or equivalently that the u_n's are uniformly integrable, where $u_n = y_n v_n$ for $n = 1, 2, \ldots$. It follows from (7.4) and (7.5) that

$$\int_{\partial D} u_n \, dm = \int_{\partial D} |u_n| \, dm \quad \text{for } n = 1, 2, \ldots .$$

Thus $u_n \geqslant 0$ for $n = 1, 2, \ldots$.

Suppose that the functions u_n's are not uniformly integrable. Then using a standard gliding hump procedure (as in [**K-P**, Lemma 2]) we construct a subsequence (u_k') of (u_n) and a sequence (e_k) of mutually disjoint closed subsets of ∂D such that for some $\delta > 0$

(7.7) $$\int_{e_k} u_k' \, dm > \delta \quad \text{and} \quad u_k' \geqslant 0 \qquad \text{for } k = 1, 2, \ldots,$$

(7.8) the functions $\chi_{\partial D \setminus e_k} \cdot u_k'$ are uniformly integrable.

Since $\sup_n \|u_n\|_1 = \sup_n \|v_n\|_1 < \infty$, we can choose the sequence (u_k') so that

(7.9) there exists a $\mu \in C(\partial D)^*$ which is the limit of the sequence (u_k') in the $\sigma(C(\partial D)^*, C(\partial D))$-topology, i.e., $\lim_k \int_{\partial D} x u_k' \, dm = \int_{\partial D} x \, d\mu$ for $x \in C(\partial D)$.

Let $\mu = u \cdot m + \nu$ where $u \in L^1$ and ν is singular with respect to m. By (7.7), $u \geqslant 0$, $\nu \geqslant 0$ and $\|\mu\| \geqslant \delta > 0$.

We shall denote by (y_k') and (v_k') subsequences of (y_n) and (v_n), respectively corresponding to the subsequence (u_k') of (u_n); we have $u_k' = y_k' v_k'$ for $k = 1, 2, \ldots$. The crucial part of the proof is, exploiting the fact that $\mu > 0$, to construct in A a w.u.s. sequence (x_s) and a subsequence (u_{k_s}') so that the integrals $\int x_s u_{k_{s+1}}' \, dm = \int (x_s y_{k_{s+1}}') v_{k_{s+1}}' \, dm$ stay uniformly away from zero. This will contradict (4), because (x_s) w.u.s. yields $(x_s y_{k_{s+1}}')$ w.u.s.

We shall consider two cases separately. The first one heavily depends on Havin's lemma (Proposition 6.1). In the second one we use only the fact that the annihilator of A is norm separable.

Case I: $\nu = 0$. Then $a = \int_{\partial D} u \, dm = \|\mu\| > \delta$. Let $f_k = f_{e_k}$ and $g_k = g_{e_k}$ for $k = 1, 2, \ldots$, where the functions f_{e_k} and g_{e_k} are constructed in Havin's lemma (Proposition 6.1). Since e_k are closed, f_k and g_k belong to A. Clearly (6.1)–(6.3) yield

(7.10) $$\lim_k \int_{\partial D} h g_k \, dm = \int_{\partial D} h dm \quad \text{for } h \in L^1,$$

(7.11) $$\lim_k \sup_{z \in e_k} |f_k(z) - 1| = 0.$$

Let us fix $\eta > 0$ so that $\delta - a\eta > \delta/2$. Using (7.9)–(7.11) we construct an increasing sequence of the indices $k_j < k_2 < \cdots$ so that if $G_0 = 1$ and $G_s = \Pi_{j=1}^s g_{k_j}$ for $s \geqslant 1$, then for $s = 1, 2, \ldots$

$$\left| \int_{\partial D} G_s u \, dm - \int_{\partial D} G_{s-1} u \, dm \right| < 2^{-s-1} a\eta, \tag{7.12}$$

$$\left| \int_{\partial D} G_s u \, dm - \int_{\partial D} G_s u'_{k_{s+1}} \, dm \right| < 2^{-s-2} a\eta, \tag{7.13}$$

$$|f_{k_{s+1}}(z) - 1| < 2^{-s} a\eta \quad \text{for } z \in e_{k_{s+1}}. \tag{7.14}$$

Now we define a w.u.s. sequence (x_s) in A by

$$x_s = G_s f_{k_{s+1}} \quad \text{for } s = 1, 2, \ldots .$$

The sequence (x_s) is w.u.s. because, by (6.1),

$$\sum_{j=1}^{s} |x_s(z)| \leqslant \prod_{j=1}^{s+1} (|f_{k_j}(z)| + |g_{k_j}(z)|) = 1 \quad \text{for } z \in \partial D \text{ and for } s = 1, 2, \ldots .$$

Combining (7.12) with (7.13) we get

$$\begin{aligned} \left| \int_{\partial D} G_s u'_{k_{s+1}} \, dm \right| &\geqslant \left| \int_{\partial D} G_s u \, dm \right| - 2^{-s-2} a\eta \\ &\geqslant \int_{\partial D} u \, dm - \sum_{j=1}^{s} \left| \int_{\partial D} G_j u \, dm - \int_{\partial D} G_{j-1} u \, dm \right| - 2^{-s-2} a\eta \\ &\geqslant a - \sum_{j=1}^{s} 2^{-j-2} a\eta - 2^{-s-2} a\eta > a(1 - \eta). \end{aligned} \tag{7.15}$$

Clearly $\|G_s\|_\infty \leqslant 1$ (by (6.1)), $\sup_{z \in e_{k_{s+1}}} |f_{k_{s+1}}(z) - 1| \longrightarrow 0$ as $s \longrightarrow \infty$ (by 7.11), and $\sup_k \|u'_k\|_1 \leqslant \sup_n \|v_n\| = M < \infty$. Thus

$$\lim_{s=\infty} \int_{e_{k_{s+1}}} |1 - f_{k_{s+1}}| \, |G_s| u'_{k_{s+1}} \, dm = 0$$

and (by (6.2)) $\lim_{s=\infty} \int_{\partial D \setminus e_{k_{s+1}}} |1 - f_{k_{s+1}}| \, dm = 1$. Hence, by (7.8) and the relation $\lim_{s=\infty} \int_{\partial D} u'_{k_{s+1}} \, dm = \int_{\partial D} u \, dm = a$, we get

$$\begin{aligned} \limsup_{s=\infty} \int_{\partial D} |1 - f_{k_{s+1}}| \, |G_s| u'_{k_{s+1}} \, dm &= \limsup_{s=\infty} \int_{\partial D \setminus e_{k_{s+1}}} |1 - f_{k_{s+1}}| \, |G_s| u'_{k_{s+1}} \, dm \\ &\leqslant \limsup_{s=\infty} \int_{\partial D \setminus e_{k_{s+1}}} u'_{k_{s+1}} \, dm \\ &\leqslant a - \liminf_{s=\infty} \int_{e_{k_{s+1}}} u'_{k_{s+1}} \, dm \\ &\leqslant a - \delta. \end{aligned} \tag{7.16}$$

Combining (7.15) with (7.16) we get

$$\liminf_{s=\infty} \left| \int_{\partial D} x_s u'_{k_{s+1}} \, dm \right|$$

$$\geq \liminf_{s=\infty} \left| \int_{\partial D} G_s u'_{k_{s+1}} \, dm \right| - \limsup_{s=\infty} \int_{\partial D} |1 - f_{k_{s+1}}| \, |G_s| u'_{k_{s+1}} \, dm \geq \frac{\delta}{2}.$$

Finally we put

$$\varphi_s = y'_{k_{s+1}} \cdot x_s \quad \text{for } s = 1, 2, \ldots .$$

Since (x_s) is w.u.s. and $\|y'_k\| = 1$ for $k = 1, 2, \ldots$, the sequence (φ_s) is w.u.s. Clearly we have

$$\int_{\partial D} \varphi_s v'_{k_{s+1}} \, dm = \int_{\partial D} x_s u'_{k_{s+1}} \, dm \quad \text{for } s = 1, 2, \ldots .$$

Thus

$$\limsup_{s=\infty} \sup_{\{w+H^1_0\} \in W} \left| \int_{\partial D} \varphi_s w \, dm \right| \geq \liminf_{s=\infty} \left| \int_{\partial D} x_s u'_{k_{s+1}} \, dm \right| \geq \frac{\delta}{2}.$$

This contradicts (4) and completes the proof in Case I.

Case II: $\nu > 0$. Since the measure ν is regular and singular with respect to m, there is a closed set $e \subset \partial D$ with $m(e) = 0$ such that $\nu(e) \geq 7\|\nu\|/8 = 7a/8$. Using Bishop's general Rudin-Carleson theorem (cf. §2, Theorem 2.1) we construct inductively a sequence $(f_n)_{n \geq 0}$ in A and a sequence $(e_n)_{n \geq 1}$ of closed subsets of ∂D such that

$$f_0 = 0, \quad f_1 = 1, \quad e_1 = \partial D, \qquad e_2 = \{z \in \partial D: |z - w| \leq 2^{-1} \text{ for } w \in e\},$$

$$f_n(z) = \|f_n\| = 1 \quad \text{for } z \in e, \qquad |f_n(z)| < 2^{-n} \quad \text{for } z \notin e_n$$

(7.17)
$$e_{n+1} = e_n \cap \left\{ z \in \partial D: \sup_{w \in e} |z - w| \leq 2^{-n} \text{ and } \sum_{j=2}^{n} |f_j(z) - f_{j-1}(z)| \leq 2^{-n} \right\}$$

$$(n = 2, 3, \ldots).$$

Observe that

$$S(z) = \sum_{j=1}^{\infty} |f_j(z) - f_{j-1}(z)| \leq 3 \quad \text{for } z \in \partial D.$$

Indeed, if $z \in e$, then $S(z) = 1$. If $z \notin e$, then $z \in e_{k-1} \setminus e_k$ for some $k = 2, 3, \ldots$; if $k = 2$, then

$$S(z) \leq 2 \sum_{j=1}^{\infty} |f_j(z)| \leq 2 + 2 \sum_{n=2}^{\infty} 2^{-n} \leq 3;$$

if $k > 2$, then

$$S(z) = \sum_{j=1}^{k-2} |f_j(z) - f_{j-1}(z)| + |f_{k-1}(z)| + 2 \sum_{j=k}^{\infty} |f_j(z)|$$

$$\leq 2^{-k+2} + 1 + 2 \sum_{j=k}^{\infty} 2^{-j} < 3.$$

Clearly $\bigcap_{j=1}^{\infty} e_j = e$. Hence using (7.9) and (7.17) we construct inductively increasing sequences of the indices (n_s) and (k_s), so that for $s = 1, 2, \ldots$,

$$\left| \int_{\partial D} f_{n_s} u'_{k_j}\, dm \right| \leqslant 2^{-s} a, \quad \text{for } j \leqslant s,$$

$$\left| \int_{\partial D} f_{n_s} u\, dm \right| \leqslant 2^{-s-2} a,$$

$$\left| \int_{\partial D} f_{n_s} u'_{k_{s+1}}\, dm - \int_{\partial D} f_{n_s}\, d\mu \right| \leqslant 2^{-s-2} a.$$

Let us put $n_0 = 0$ and $x_s = f_{n_s} - f_{n_{s+1}}$ $(s = 1, 2, \ldots)$. Then (x_s) is w.u.s. because, for $z \in \partial D$,

$$\sum_{s=1}^{\infty} |x_s(z)| \leqslant \sum_{s=1}^{\infty} |f_{n_s}(z) - f_{n_{s-1}}(z)| \leqslant \sum_{j=1}^{\infty} |f_j(z) - f_{j-1}(z)| \leqslant 3.$$

Furthermore, for $s = 1, 2, \ldots$,

$$\begin{aligned} \left| \int_{\partial D} x_s u'_{k_{s+1}}\, dm \right| &\geqslant \left| \int_{\partial D} f_{n_s} u'_{k_{s+1}}\, dm \right| - \left| \int_{\partial D} f_{n_{s+1}} u'_{k_{s+1}}\, dm \right| \\ &\geqslant \left| \int_{\partial D} f_{n_s}\, d\mu \right| - 2^{-s-2} a - 2^{-s-1} a \\ &\geqslant \left| \int_{\partial D} f_{n_s}\, d\nu \right| - \left| \int_{\partial D} f_{n_s} u\, dm \right| - 2^{-s-2} a - 2^{-s-1} a \\ &\geqslant \left| \int_{e} f_{n_s}\, d\nu \right| - \left| \int_{\partial D \setminus e} f_{n_s}\, d\nu \right| - 2^{-s-2} a - 2^{-s-2} a - 2^{-s-1} a \\ &\geqslant \nu(e) - \nu(\partial D \setminus e) - 2^{-s} a \geqslant \tfrac{1}{4} a. \end{aligned}$$

Now we put $\varphi_s = x_s y'_{k_{s+1}}$ for $s = 1, 2, \ldots$ and we complete the proof as in Case I. □

COROLLARY 7.1. *The assertion of Theorem* 7.1 *remains valid if* L^1/H_0^1 *is replaced by* A^*, L^1 *by* $C(\partial D)^*$ *and the condition* (4) *by*

(4a) $\lim_n (\sup_{\{\mu + H_0^1\} \in W} |\int_{\partial D} \varphi_n\, d\mu|) = 0$ *for every w.u.s. sequence* (φ_n) *in* A.

PROOF. To derive the implication (4a) ⇒ (1) from Theorem 7.1, denote by p_1 and p_2 the natural projections from $A^* = C(\partial D)^*/H_0^1 = L^1/H_0^1 \oplus_1 V_{\text{sing}}$ onto L^1/H_0^1 and V_{sing} respectively, and let $\tau: C(\partial D)^*/H_0^1 \longrightarrow C(\partial D)^*$ and $\tau_L: L^1/H_0^1 \longrightarrow L^1$ to be the nearest point cross-sections. Now if $W \subset A^*$ satisfies (4a), then both of the sets $p_1(W)$ and $p_2(W)$ have the same property. For subsets of L^1/H_0^1, (4a) is equivalent to (4). Hence the weak closure of $\tau(p_1(W)) = \tau_L(p_1(W))$ is weakly compact. Furthermore, for subsets of V_{sing}, (4a) combined with the general Rudin-Carleson theorem and the identity $\tau p_2 = p_2$ yields $\lim_n(\sup_{\mu \in p_2(W)} |\int_{\partial D} \varphi_n\, d\mu|) = 0$ for every w.u.s. sequence (φ_n) in $C(\partial D)$. Thus, by a result of [P9], the weak closure of the set $\tau(p_2(W)) = p_2(W)$ is weakly compact. Since

$$\tau(W) \subset \tau(p_1(W) \oplus_1 p_2(W)) = \tau_L(p_1(W)) \oplus_1 p_2(W),$$

we conclude that the weak closure of $\tau(W)$ is weakly compact. □

COROLLARY 7.2. *A weak closure of a uniformly bounded set* $W \subset L^1/H_0^1$ *is weakly compact iff there exists a function* $\epsilon \longrightarrow K(\epsilon)$ *for* $\epsilon > 0$ *such that for every* $\{w + H_0^1\} \in W$ *there exists a* $g \in H^\infty$ *with* $\|g\|_\infty \leqslant K(\epsilon)$ *such that* $\|q(w - g)\|_{L^1/H_0^1} < \epsilon$ ($q\colon L^1 \longrightarrow L^1/H_0^1$ *denotes the quotient map*).

Hint. Use Theorem 7.1 and the uniform integrability criterion for weak compactness of subsets of L^1. □

8. Weakly Compact Operators from A, L^1/H_0^1 and A^* and Complemented Subspaces of These Spaces

In this section we reap the benefits of Theorem 7.1. We establish further similarities between A and $C(\partial D)$, and between L^1 and L^1/H_0^1. We show that A, A^* and L^1/H_0^1 are weakly complete and have the Dunford-Pettis property. We also show that weakly compact operators from A into an arbitrary Banach space are characterized by their behavior on subspaces isomorphic to c_0, exactly in the same way as the weakly compact operators from a $C(S)$-space. In the second part of the section, we deal with complemented subspaces of A and L^1/H_0^1. We show, in particular, that L^1/H_0^1 does not contain any complemented subspace isomorphic to L^1. At the end of the section we discuss various related open problems.

Recall that a Banach space X is *weakly complete* if every weak Cauchy sequence in X converges weakly to some element of X. We say that X has the *Dunford-Pettis property* if every weakly compact operator from X into arbitrary Banach space takes weak Cauchy sequences into norm convergent sequences.

The proofs of the Corollaries 8.1–8.5 below are, "modulo Theorem 7.1", easy modifications of analogous results for $C(S)$-spaces and L^1-spaces.

COROLLARY 8.1. (a) *The spaces L^1/H_0^1 and A^* are weakly complete.*
(b) *The space L^1/H_0^1, A^* and A have the Dunford-Pettis property.*

PROOF. (a) follows immediately from the equivalence of the conditions (2) and (3) of Theorem 7.1 (resp. Corollary 7.1). To prove (b) for L^1/H_0^1, resp. for A^*, observe that the equivalence of conditions (1) and (3) of Theorem 7.1 (resp. Corollary 7.1) implies that every weak Cauchy sequence in L^1/H_0^1 (resp. in $C(\partial D)^*$) is the image under a quotient map of a weak Cauchy sequence in L^1 (resp. in $C(\partial D)^*$). Now we use the fact that L^1 (resp. $[C(\partial D)]^*$) has the Dunford-Pettis property (cf. **[D-SI**, Chapter VI**]**). The assertion (b) for A follows from Grothendieck's observation that if X^* has the Dunford-Pettis property then X does (cf. **[Gr2]**, **[P9]**). □

COROLLARY 8.2. *If E is one of the spaces A, A^*, L^1/H_0^1 and $T: E \longrightarrow E$ a weakly compact operator, then T^2 is compact.* □

COROLLARY 8.3. *Let Y be a Banach space and let $T: A \longrightarrow Y$ be a bounded linear operator. Then the following conditions are equivalent:*

(i) *T is weakly compact.*

(ii) *T takes weak Cauchy sequences into convergent sequences.*

(iii) *T restricted to any isomorph of c_0 is not invertible.*

(iv) *T takes every w.u.s. sequence in A into a sequence which converges to zero.*

PROOF. (i) $\Rightarrow$ (ii). Apply Corollary 8.1(b) for A.

(ii) $\Rightarrow$ (iii). Obvious, because there are weak Cauchy sequences in c_0 which do not converge in norm.

(iii) $\Rightarrow$ (iv). Note that a bounded linear operator takes w.u.s. sequences into w.u.s. ones and use the result of **[B-P]** which says that if (x_n) is a w.u.s. sequence in a Banach space X which does not converge to zero in norm, then there is an isomorphic embedding from c_0 into X which takes the unit vectors of c_0 onto a subsequence of the sequence (x_n).

(iv) $\Rightarrow$ (i). Observe that if T satisfies (iv), then $W = T^*(B_A)$ satisfies the condition (4a) of Corollary 7.1. Hence, by this corollary, the weak closure of W in $C(\partial D)^*$ is weakly compact. Hence T^* is weakly compact and therefore T is weakly compact (cf. **[D-SI**, Chapter VI]). □

COROLLARY 8.4. *Let W be a bounded subset of A^* (resp. L^1/H_0^1) whose weak closure is not weakly compact. Then there is a sequence (x_n^*) of elements of W which is equivalent to the unit vector basis of l^1 and spans a complemented subspace of A^* (resp. L^1/H_0^1). More precisely, there are operators $T\colon l^1 \longrightarrow [C(\partial D)]^*$ (resp. $T\colon l^1 \longrightarrow L^1$) and $S\colon A^* \longrightarrow l^1$ (resp. $S\colon L^1/H_0^1 \longrightarrow l^1$) such that $SqT = id_{l^1}$ and $qTe_n = x_n^*$ for $n = 1, 2, \ldots$ where $q\colon [C(\partial D)]^* \longrightarrow A^*$ (resp. $q\colon L^1 \longrightarrow L^1/H_0^1$) denotes the restriction (the quotient) map and e_n denotes the nth unit vector of l^1.*

PROOF. The assumption on W combined with Corollary 7.1 yields the existence of a w.u.s. sequence (φ_s) in A and a sequence (y_s^*) in W such that $|y_s^*(\varphi_s)| \geqslant 1$ for $s = 1, 2, \ldots$. Let

$$M = \max\left(\sup_s \|y_s^*\|, \sup_{\|y^*\|=1} \sum_s |y^*(\varphi_s)|\right).$$

Clearly $M < +\infty$. Passing to a subsequence, if necessary, we may also assume that $\Sigma_{t=s+1}^{\infty} |y_s^*(\varphi_t)| < 2^{-s-3}$ for $s = 1, 2, \ldots$. Next we define inductively a sequence of the indices $(k(n))_{n=0}^{\infty}$ and a sequence of infinite subsets of the integers $(N_n)_{n=0}^{\infty}$ such that N_0 – the set of all positive integers, $k_0 = 0$,

$$N_{n-1} \supset N_n \quad \text{and} \quad k(n) \in N_{n-1} \qquad (n = 1, 2, \ldots), \tag{8.1}$$

$$\text{if } j \in N_n, \text{ then } j > k(n) \text{ and } |x_j^*(\varphi_{k(n)})| < 2^{-n-3} \qquad (n = 1, 2, \ldots). \tag{8.2}$$

Suppose that for some $q \geqslant 0$ and for $0 \leqslant r \leqslant q$ the indices $k(r)$ and the infinite sets $N(r)$ have been constructed to satisfy (8.1) and (8.2). Fix an integer $m > M2^{q+4}$ and let A_q be a set of distinct m indices in N_q which are $> k(q)$. Then for each $j = 1, 2, \ldots$ there is an index $t(j) \in A_q$ such that $|x_j^*(\varphi_{t(j)})| < 2^{-q-4}$, because otherwise $\Sigma_{t=1}^{\infty} |x_j^*(\varphi_t)| \geqslant \Sigma_{t\in A_q} |x_j^*(\varphi_t)| > M$ which leads to a contradiction. Since N_q is an infinite set and A_q is a finite one, we infer that there is an infinite subset N_{q+1} of the set $N_q \cap \{k(q) + 1, k(q) + 2, \ldots\}$ and an index $k(q + 1) \in A_q$ such that if $j \in N_{q+1}$, then $|x_j^*(\varphi_{k(q+1)})| < 2^{-q-4}$. This completes the induction.

Now put $\psi_n = \varphi_{k(n)}[y^*_{k(n)}(\varphi_{k(n)})]^{-1}$ and $x^*_n = y^*_{k(n)}$ $(n = 1, 2, \ldots)$ and define S: $A^* \longrightarrow l^1$ by $S(x^*) = (x^*(\psi_n))$ for $x^* \in A^*$. A straightforward computation shows that for every finite sequence of scalars $c_1, c_2, \ldots, c_m$ $(m = 1, 2, \ldots)$

$$\left(1 + \frac{1}{4}\right)\sum_{k=1}^{m} |c_j| \geqslant \left\|\sum_{j=1}^{m} c_j S(x_j^*)\right\| \geqslant \left(1 - \frac{1}{4}\right)\sum_{j=1}^{m} |c_j|.$$

Applying a standard stability argument (cf. **[B-P]**) we construct an isomorphism V: $l^1 \xrightarrow[\text{onto}]{} l^1$ such that $VS(x_n^*) = e_n$ for $n = 1, 2, \ldots$. Finally define T: $l^1 \longrightarrow [C(\partial D)]^*$ by $T(e_n) = \tau(x_n^*)$ for $n = 1, 2, \ldots$ where τ: $A^* \longrightarrow [C(\partial D)]^*$ is the nearest point cross-section. The proof for L^1/H^1_0 is the same. □

Next we shall discuss properties of complemented subspaces of A, L^1/H^1_0 and H^*. Our first corollary generalizes results for $C(S)$ and $L^1(\nu)$ spaces (cf. **[P1]** and **[R2]**); it shows in particular that A, L^1/H^1_0 and A^* do not have complemented infinite dimensional reflexive subspaces.

COROLLARY 8.5. *Let X be an infinite dimensional Banach space.*

(a) *If X is complemented in A, then X contains an isomorph of c_0.*

(b) *If X is complemented in A and X^* is not separable, then X contains an isomorph of $C(\partial D)$.*

(c) *If X is complemented either in A^* or in L^1/H^1_0, then X contains an isomorph of l^1.*

(d) *If X is a separable dual and is complemented either in A^* or in L^1/H^1_0 then weak and norm convergence of sequences in X coincide.*

(e) *If X has a G-L l.u.st and X is complemented in L^1/H^1_0 then weak and norm convergence of sequences in X coincide.*

PROOF. (a) Apply Corollaries 8.1(b) and 8.2. (c) Combine Corollary 8.1(b) with Corollaries 8.2 and 8.4. (d) Use Corollary 8.1(b) combined with a simple observation that a complemented subspace of a Banach space with the Dunford-Pettis property has the Dunford-Pettis property, and apply a result of Grothendieck (cf. **[Gr2]** and **[P4]**) that if a separable dual space has the Dunford-Pettis property, then weak and norm convergence of sequences coincide in the space.

PROOF OF (b). Let Y be a complementary subspace to X in A. Since A/Y is isomorphic to X, the hypothesis implies that $(A/Y)^*$ is not separable. Thus $(C(\partial D)/Y)^*$ is not separable and, by a result of Rosenthal **[R2]**, $C(\partial D)/Y$ contains a subspace isomorphic to $C(\partial D)$. By a result of **[L-P2]** (cf. also **[R2]**), this implies that A/Y contains a subspace isomorphic to $C(\partial D)$ because, by the F. and M. Riesz theorem the space $(C(\partial D)/Y)/(A/Y)$ has a separable dual. This completes the proof because A/Y is isomorphic to X.

REMARK. In fact (b) is true for every Banach space which is isomorphic to a subspace of a $C(S)$-space with separable annihilator, provided S is an uncountable compact metric space.

PROOF OF (e). The argument is similar to that of Theorem 4.2(v). We consider the diagram

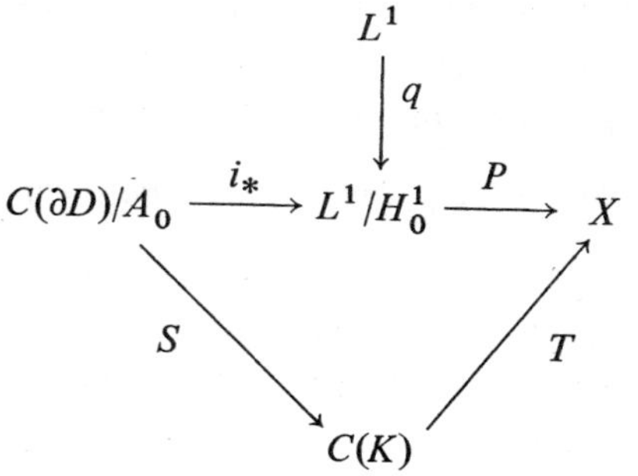

where $P\colon L^1/H_0^1 \xrightarrow[\text{onto}]{} X$ is a projection, $q\colon L^1 \longrightarrow L^1/H_0^1$ is the quotient map, and $i_*\colon C(\partial D)/A_0 \longrightarrow L^1/H_0^1$ is defined by $i_*(\{f + A_0\}) = \{f + H_0^1\}$ for $f \in C(\partial D)$. To define operators S and T we first note that $(i_*)^* = i_{m,1}^{H^\infty}\colon H^\infty \longrightarrow H^1$ is an absolutely summing operator. Hence $(Pi_*)^*$ is absolutely summing with values in the separable dual space H^1. Since X has G-L l.u.st, Theorem 4.1 combined with Remark R.III in §4 yields that $i_*^*P^*$ is L^1-factorable. Hence $(Pi_*)^* = VU$ for some $L^1(\nu)$-space and operators $U\colon X^* \longrightarrow L^1(\nu)$ and $V\colon L^1(\nu) \longrightarrow H^1$. We now define $C(K)$ to be the dual of $L^1(\nu)$, S to be the restriction of V^* to $C(\partial D)/A_0$ [identified with its canonical image in $(C(\partial D)/A_0)^{**}$] and we put $T = QU^*$, where Q is a projection from X^{**} onto X. Note that L^1/H_0^1 is complemented in the dual Banach space A^* and therefore every complemented subspace of L^1/H_0^1 is complemented in the second dual of the subspace.

Next we observe that $Pi_* = TS$ is a compact operator. Indeed note that $T\colon C(S) \longrightarrow X$ takes weakly Cauchy sequences into convergent sequences because $C(S)$ has the Dunford-Pettis property (cf. **[D-SI**, Chapter VI**]**) and, by a result of **[P1]**, T is weakly compact. (The range of T, being contained in the weakly complete space L^1/H_0^1, does not contain subspaces isomorphic to c_0.) Furthermore every sequence in the unit ball $B_{C(\partial D)/A_0}$ has a weak Cauchy subsequence because $(C(\partial D)/A_0)^* = H^1$ is separable. Hence $TS(B_{C(\partial D)/A_0})$ is a totally bounded subset of X.

Now suppose that (x_n) is a sequence which is weakly convergent to zero in X. By Theorem 7.1, there is a sequence (f_n) in L^1 which consists of uniformly bounded and uniformly integrable functions, and $q(f_n) = x_n$ for $n = 1, 2, \ldots$. The properties of (f_n) combined with the density in L^1 of all the continuous functions yield that for every $\epsilon > 0$ there is an $M = M(\epsilon) < +\infty$ and a sequence (g_n) in $C(\partial D)$ such that $\|g_n\|_\infty \leqslant M$ and $\|g_n - f_n\|_1 < \epsilon$ for $n = 1, 2, \ldots$. Therefore the elements of the sequence $(Pi_*(\{g_n + A_0\}))$ form a totally bounded subset of X because Pi_* is compact and $\sup_n \|\{g_n + A_0\}\| \leqslant \sup_n \|g_n\| \leqslant M$. On the other hand we have, for $n = 1, 2, \ldots$,

$$\|Pi_*(\{g_n + A_0\}) - x_n\| = \|P(\{g_n + H_0^1\}) - Pq(f_n)\|$$

$$= \|P\|\,\|g_n - f_n\|_1 \leqslant \|P\|\epsilon.$$

Since $\epsilon > 0$ has been chosen arbitrarily, we infer that the set $\bigcup_n \{x_n\}$ is totally bounded in X. Hence the weakly convergent sequence (x_n) converges in norm. □

Our next corollary which is an immediate consequence of Corollary 8.5 shows that L^1 does not have "the strong primary property" analogous to the property of $C([0, 1])$ discovered in **[L-P2]**. Note that H_0^1, being separable dual, does not contain isomorphs of L^1.

COROLLARY 8.6. *The space L^1/H_0^1 does not contain a complemented isomorph of L^1.*

Problem 8.1. Does there exist a subspace of L^1/H_0^1 which is isomorphic to L^1?

Notes and remarks to §§7 *and* 8. In 1967 Piranian, Shields and Wells **[P-S-W]** raised the question whether L^1/H_0^1 is weakly complete in the following form.

Given a bounded sequence (g_n) in L^1 such that the limit $\lim_n \int_{\partial D} fg_n\, dm = \Phi(f)$ exists for every $f \in H^\infty$, does there exist a $g \in L^1$ such that $\Phi(f) = \int_{\partial D} fg\, dm$ for $f \in H^\infty$? Kahane **[Kh2]** observed that there is a g in question such that $\Phi(f) = \int_{\partial D} fg\, dm$ for $f \in A$.

The weak completeness of L^1/H_0^1 and therefore of A^* (our Corollary 8.1(a)) was proved by Mooney **[Moo]** and independently by Havin **[Hv1]**. Applying the result of **[A-L]** on peak sets in H^∞, Amar **[A]** adopted the method of **[Kh2]** to get an elegant proof of the Mooney-Havin result. Chaumat **[Chm]** proved that L^1/H_0^1 and therefore A^* and A have the Dunford-Pettis property. A different proof of this result and the "liftability" of weakly compact sets in L^1/H_0^1 to weakly compact sets in L^1 was obtained by Cnop and Delbaen **[C-D]**. Chaumat's approach generalizes to the case of a predual of a uniform algebra in which an abstract analogue of the Amar-Lederer lemma (our Proposition 6.2) holds, while the Delbaen and Cnop method generalizes to uniform algebras satisfying some conditions imposed on sets of representing measures (cf. **[C-D]** and **[D2]**). Theorem 7.1 in the present form and most of the corollaries in §8 have been obtained independently by Delbaen **[D1]** and Kisliakov **[Kis2]**. Wojtaszczyk **[W2]** recently generalized Theorem 7.1 to certain planar uniform algebras. The proof of Theorem 7.1 presented in the text follows with some modifications the argument in **[Kis2]**. The construction in Case II goes back to Chaumat **[Chm]**; our argument is different. Corollary 8.5(e) and Corollary 8.6 are due to W. B. Johnson and are published here with his permission.

Several questions remain open. Very little is known about the nature of the dual of H^∞. Kisliakov in **[Kis2]** claims that the assertion of Theorem 7.1 is also true in the case of H^∞. Unfortunately his argument contains a gap which he has recognized and communicated to me.

Let us consider the following properties of a Banach space X:

(a) the Dunford-Pettis property of X,

(b) the Dunford-Pettis property of the dual of X,

(c) the weak completeness of the dual of X,

(d) a set $W \subset X^*$ is weakly compact if it is weakly closed and $\sup_{w^* \in W} |w^*(f_n)| = 0$ for every w.u.s. sequence (f_n) in X,

(e) every sequence in X^* which converges in the $\sigma(X^*, X)$-topology converges weakly,

(f) every complemented subspace of X is either finite dimensional or contains l^∞.

Problem 8.2. Does H^∞ have the properties (a)–(f)?

Problem 8.3. Let S be a compact metric space and let X be a subspace of $C(S)$ with a norm separable annihilator. Does X have the properties (a)–(d)? How about the special case in which it is assumed in addition that X is a uniform algebra on S?

Problem 8.4. Does the n-disc algebra $A(D^n)$ (resp. the n-ball $A(B_n)$) have the properties (a)–(d) for $n \geqslant 2$? (For the definition of $A(D^n)$ and $A(B_n)$ see §11.)

Problem 8.5. What translation invariant subspaces of $C(\partial D_n)$ have some of the properties (a)–(d)?

Our next problem concerns complemented subspaces of A and L^1/H^1_0.

Problem 8.6. (a) Let E be an infinite dimensional complemented subspace of A. Assume that E has an unconditional basis (resp. E has a G-L l.u.st). Is E isomorphic to c_0 (resp. is E isomorphic to a complemented subspace of $C(\partial D)$)?

(b) Let E be an infinite dimensional complemented subspace of L^1/H^1_0. Assume that E has one of the following properties:

(1) E has an unconditional basis,

(2) E has G-L l.u.st,

(3) E is a separable dual,

(4) E has the Radon-Nikodym property.

Is E isomorphic to l^1?

Our last problem is closely related to the fact that every weakly compact set in L^1/H^1_0 can be lifted to a weakly compact set in L^1 (cf. Theorem 7.1). It is also related to Problem 3.1.

Problem 8.7. Let E be a reflexive subspace of L^1/H^1_0. Does there exist a reflexive subspace E_1 of L^1 such that the quotient map $q\colon L^1 \longrightarrow L^1/H^1_0$ restricted to E_1 is an isomorphism from E_1 onto E? How about the special case when E is isomorphic to a Hilbert space?

ADDED IN PROOF. 1° Delbaen **[De3]** generalized Theorem 7.1 for uniform algebras such that every linear multiplicative functional has a weakly compact set of representing measures and there is no nontrivial measure which is singular with respect to all the representing measures.

2° Kisliakov announces in **[Kis4]** that if E is a reflexive subspace of A^* then there exists a bounded linear operator $T\colon A^* \longrightarrow A^*$ such that $T|E$ is an isomorphism and there exists an $E_1 \subset [C(\partial D)]^*$ such that the quotient map maps isomorphically E_1 onto E.

3° The answers on Problems 8.6(b)(3) and 8.6(b)(4) are negative (Delbaen private communication).

9. Complementation of Finite Dimensional Subspaces in A, L^1/H_0^1 and H^∞

The technique of Theorem 7.1 provides no information on the degree of complementation of finite dimensional subspaces of A (resp. L^1/H_0^1). For instance Corollary 8.5(a) obviously implies that infinite dimensional hilbertian subspaces are uncomplemented in A but it yields no estimate of the growth of the norm of the best projection onto n-dimensional Hilbert subspaces of A. Our approach to handle the above problem and similar questions concerning the "local" structure of A and L^1/H_0^1 is based upon Theorem 2.4. This theorem and the duality relation between A, L^1/H_0^1, and H^∞ are the only facts on spaces of analytic functions which are used in this section, the rest is a good example of "purely formal" technique of the local theory of Banach spaces and the theory of absolutely summing operators and related Banach ideals.

We shall show (cf. Theorems 9.1 and 9.2) that A, L^1/H_0^1, H^∞ do not contain nicely complemented l_n^p's for $1 < p < \infty$; A and H^∞ do not contain nicely complemented l_n^1's, equivalently L^1/H_0^1 does not contain l_n^∞ uniformly (cf. Theorem 4.2(iv) for definition).

We begin with introducing two concepts; the first one plays an important role in the local theory of Banach spaces, the second is a specific invariant suggested by Theorem 2.4.

DEFINITION 9.1. Let X and Y be Banach spaces; a pair (S, T) of linear operators S: $X \longrightarrow Y$, T: $Y \longrightarrow X$ with $TS = \mathrm{id}_X$ is said to be a *factorization of* X *through* Y. The *factorization constant of* X through Y is the quantity

$$\gamma_Y(X) = \begin{cases} \inf \|S\|\,\|T\| & \text{if there are factorizations of } X \text{ through } Y, \\ +\infty & \text{otherwise.} \end{cases}$$

Here "inf" is extended over all factorization of X through Y.

DEFINITION 9.2. Let X be a Banach space and let $1 < p < \infty$. The $i_p - \pi_p$ *ratio* of X is the quantity

$$k_p(X) = \sup i_p(T)$$

where the supremum extends over all operators T from X into arbitrary Banach spaces with $\pi_p(T) = 1$.

The concepts defined above are related between themselves and with projections. We have

PROPOSITION 9.1. *Let X, Y and X_1 be Banach spaces and let $1 < p < \infty$. Then*

(a) $k_p(X) \leqslant \gamma_Y(X) k_p(Y)$;

(b) *if* X_1 *is a subspace of* Y *and* $P: Y \xrightarrow[onto]{} X_1$ *a projection then*

$$\|P\| \geq \gamma_Y(X)/\gamma_{X_1}(X)$$

provided $\min(\gamma_Y(X), \gamma_{X_1}(X)) < +\infty$;

(c) *if* X *is finite dimensional and* X_1 *is isomorphic to* X *then* $\gamma_{X_1}(X) = d(X, X_1)$ *where* $d(X, X_1) = \inf\{\|S\|\|S^{-1}\| | S: X \xrightarrow[onto]{} X_1$ *isomorphism*$\}$ *is the Banach-Mazur distance between* X *and* X_1;

(d) *if* X *is finite dimensional then*

$$\gamma_Y(X) = \gamma_{Y^{**}}(X).$$

PROOF. (a) Let us fix a linear operator $U: X \longrightarrow E$ (E an arbitrary Banach space) with $\pi_p(U) = 1$ and let (S, T) be a factorization of X through Y. Then

$$i_p(U) = i_p(UTS) \leq \|S\| i_p(UT) \leq \|S\| k_p(Y)\pi_p(UT)$$
$$\leq \|S\|\, \|T\| k_p(Y)\pi_p(U).$$

Thus

$$i_p(U) \leq \inf \|S\|\, \|T\| \cdot k_p(Y)\pi_p(U) \leq \gamma_Y(X) k_p(Y)\pi_p(U)$$

which yields the desired conclusion.

We left to the reader a routine proof of (b) and (c).

(d) Clearly $\gamma_Y(X) \geq \gamma_{Y^{**}}(X)$. To prove the reverse inequality fix $\epsilon > 0$ and pick a factorization (S, T) of X through Y^{**} so that $\|S\|\,\|T\| < \gamma_{Y^{**}}(X)(1 + \epsilon)$. Let $F = S(X)$. Next note that for a finite dimensional X the space $[B(Y, X)]^{**}$ can be identified with the space $B(Y^{**}, X)$. Moreover the canonical embedding of Y into Y^{**} induces the canonical embedding of $B(Y, X)$ into $B(Y^{**}, X) = (B(Y, X))^{**}$. A simple restatement of the Goldstine theorem that canonical embedding of the unit ball of the space is weak-star dense in the unit ball of the second dual implies that there exists a $\widetilde{T} \in B(Y, X)$ such that $\|\widetilde{T}\| \leq (1 + \epsilon)\|T\|$ and $\widetilde{T}^{**}y^{**} = Ty^{**}$ for $y^{**} \in F$. Let $\widetilde{T}(y) = \Sigma_{j=1}^n y_j^*(y)x_j$ for $y \in Y$. Now we apply the Lindenstrauss-Rosenthal local reflexivity principle **[L-R]** in the improved form due to Johnson-Rosenthal-Zippin **[J-R-Z]**. Hence there exists an isomorphic embedding U: $F \longrightarrow Y$ such that $\|y^{**}\| \leq \|Uy^{**}\| \leq (1 + \epsilon)\|y^{**}\|$ for $y^{**} \in F$ and $y^{**}(y_j^*) = y_j^*(Uy^{**})$ for $y^{**} \in F$ and for $j = 1, 2, \ldots$. Then $(US, \widetilde{T})$ is a factorization of X through Y with $\|US\|\|\widetilde{T}\| \leq (1 + \epsilon)^2\|S\|\|T\| \leq \gamma_{Y^{**}}(X)(1 + \epsilon)^3$. Letting $\epsilon \longrightarrow 0$ we get $\gamma_Y(X) \leq \gamma_{Y^{**}}(X)$. □

We shall also need the following result on absolutely summing operators.

PROPOSITION 9.2 (B. MAUREY **[Mau1]**). *Let* X *be a finite dimensional Banach space, let* p *with* $1 < p < \infty$ *be fixed, and let* $k \geq 1$. *Then the following conditions are equivalent*:

(i) $i_p(T) \leq k_p\pi_p(T)$ *for every bounded linear operator* T *from* X *into arbitrary Banach space.*

(ii) *For every space* $L^{p^*}(\nu)$ *with* $p^* = p/(p - 1)$ *and every subspace* Y *of* $L^{p^*}(\nu)$ *every operator* $U: Y \longrightarrow X$ *extends to a linear operator* $\widetilde{U}: L^{p^*}(\nu) \longrightarrow X$ *with* $\|\widetilde{U}\| \leq k_p\|U\|$.

PROOF. (i) ⇒ (ii). Recall that if X is a finite dimensional Banach space and Z a reflexive one, then the space $N(X, Z)$ of all nuclear operators from X into Z is reflexive and its dual is isometrically isomorphic to the space $B(Z, X)$ of all bounded operators from Z into X; the duality is given by the trace

$$(U, V) \longrightarrow \mathrm{tr}(UV) \quad \text{for } U \in N(X, Z),\ V \in B(Z, X).$$

The condition (ii) says that the restriction operator $B(L^{p^*}(\nu), X) \longrightarrow B(Y, X)$ takes the ball in $B(L^{p^*}(\nu), X)$ of radius k_p and with the center at the origin onto the unit ball of $B(Y, X)$. Hence a standard duality argument yields that (ii) is equivalent to the following condition:

(ii)* *$n(V) \leqslant k_p n(JV)$ for every operator $V: X \longrightarrow Y$ and every closed linear subspace Y of every $L^{p^*}(\nu)$-space.*

Here $n(\cdot)$ denotes the nuclear norm of an operator and $J: Y \longrightarrow L^{p^*}(\nu)$ denotes the inclusion map. We shall show that (i) ⇒ (ii)*. Pick $V: X \longrightarrow Y$ with $n(JV) = 1$ and consider the best nuclear factorization diagram for JV, i.e., we write $JV = T\Lambda S$ where $T: X \longrightarrow l^\infty$ and $S: l^1 \longrightarrow L^{p^*}(\nu)$ are operators of norm one and $\Lambda: l^\infty \longrightarrow l^1$ is the operator of multiplication by a sequence (λ_j) with $\lambda_j \geqslant 0$ for $j = 1, 2, \ldots$ and $\Sigma\lambda_j = 1$. (The best nuclear factorization for JV does exist because X is finite dimensional and $L^{p^*}(\nu)$ is reflexive.) Clearly $\Lambda = \Lambda_{p^*}\Lambda_p$ where $\Lambda_p: l^\infty \longrightarrow l^p$ and $\Lambda_{p^*}: l^p \longrightarrow l^1$ are the operators of multiplication by the sequences $(\lambda_j^{1/p})$ and (λ_j^{1/p^*}) respectively. Let us consider the diagram

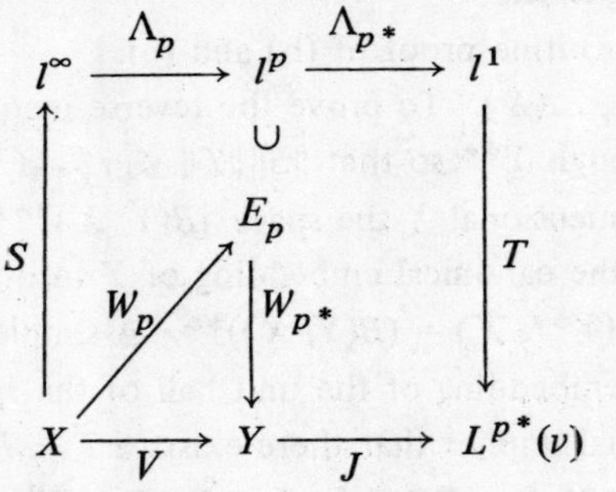

Here E_p is the closure of $\Lambda_p S(X)$ in l^p, and $W_p = \Lambda_p S$ is regarded as an operator from X into E_p. Clearly $T\Lambda_{p^*}(E_p) \subset Y$. Let W_{p^*} denote the restriction of $T\Lambda_{p^*}$ to E_p regarded as an operator to Y.

Obviously W_p is a p-absolutely summing operator with $\pi_p(W_p) = (\Sigma(\lambda_i^{1/p})^p)^{1/p} = 1$. Hence, by (i), W_p is p-integral and $i_p(W_p) \leqslant k_p\pi_p(W_p) = k_p$. By the Schwartz duality theorem (cf. [**Kw**, Theorems 1 and 2]) $T\Lambda_{p^*}$ is p^*-absolutely summing because $(T\Lambda_{p^*})^*$ is p^*-integral. Thus W_{p^*}, being restriction of a p^*-absolutely summing operator, is p^*-absolutely summing and $\pi_{p^*}(W_{p^*}) = \pi_{p^*}(T\Lambda_{p^*}) = i_{p^*}[(T\Lambda_{p^*})^*] = (\Sigma\lambda_j)^{1/p^*} = 1$. Now we use a result of Persson and Pietsch [**P-P**] or a result of Persson [**Per**] to conclude that $V = W_{p^*}W_p$ is integral and therefore nuclear (because X is finite dimensional), and

$$n(V) = i(V) \leqslant \pi_{p^*}(W_{p^*})i_p(W_p) \leqslant k_p.$$

(ii) ⇒ (i). By the Persson-Pietsch duality theory or [**P-P**], condition (i) is equivalent to the following (i)* $n_{p^*}(V) \leqslant k_p n_{p^*}(jV)$ *for every Banach space* E *and every linear operator* $V{:}E \longrightarrow X$.

Here $n_{p^*}(\,\cdot\,)$ denotes the p^*-nuclear norm of the operator and $j{:}\ X \longrightarrow l^\infty$ a fixed linear isometric embedding.

To show that (ii) $\Rightarrow$ (i)* fix $\epsilon > 0$ and an operator $V{:}\ E \longrightarrow X$ with $n_{p^*}(jV) = 1$ and consider the p^*-nuclear factorization diagram for jV

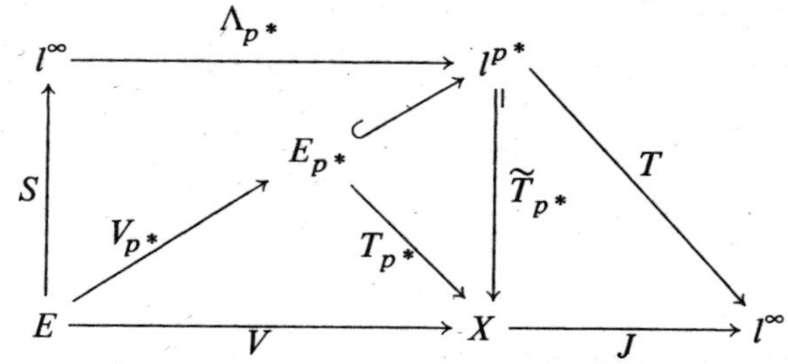

where $\Lambda_{p^*}{:}\ l^\infty \longrightarrow l^{p^*}$ is the multiplication operator by the sequence (λ_j^{1/p^*}) with $\lambda_j \geqslant 0$ and $1 \leqslant \Sigma\lambda_j \leqslant 1 + \epsilon$, and S and T are operators of norms one. Let E_{p^*} be the closure of $\Lambda_{p^*}S(X)$ in l^{p^*}, $V_{p^*} = \Lambda_{p^*}S$ regarded as an operator from E into E_{p^*}, and T_{p^*} the restriction of T to E_{p^*} regarded as an operator to X. By (ii), there is an operator $\widetilde{T}_{p^*}{:}\ l^{p^*} \longrightarrow X$ which extends T_{p^*} with $\|\widetilde{T}_{p^*}\| \leqslant k_p\|T_{p^*}\| = k_p$. Thus $V = \widetilde{T}_{p^*}\Lambda_{p^*}S$ and $n_{p^*}(V) \leqslant \|S\|\|\widetilde{T}_{p^*}\|n_{p^*}(\Lambda_{p^*}) \leqslant k_p(1+\epsilon)^{1/p^*}$. Letting $\epsilon \longrightarrow 0$, we get $n_{p^*}(V) \leqslant k_p = k_p n(V)$. □

Now we are ready to prove

THEOREM 9.1. *If* Y *is one of the spaces* A, A^*, L^1/H_0^1, H^∞ *then*

$$\gamma_Y(l_n^2) \geqslant C\frac{n^{1/2}}{\log(n+1)} \quad \textit{for } n = 1, 2, \ldots$$

where C *is an absolute constant.*

Consequently if X_1 *is an n-dimensional subspace of* Y *and* $P{:}\ Y \xrightarrow[onto]{} X_1$ *is a projection, then* $\|P\| \geqslant C(n^{1/2}/\log(n+1)) \cdot d(X_1, l_n^2)^{-1}$.

PROOF. The second part of the theorem is an obvious consequence of the first one and Proposition 9.1(b) and (c). Next observe that by Proposition 9.1(a) and the standard duality relations between A, A^*, L^1/H_0^1 and H^∞ (cf. §2) it is enough to consider the case $Y = A$ only.

Let $C_1, C_2, \ldots$ denote positive constants independent of p and n. By Theorem 2.4, $k_p(A) \leqslant C_1 p$ for $p \geqslant 2$. Next we shall show

(9.1) $$k_p(l_n^2) \geqslant C_2 n^{1/2 - 1/p} \quad \text{for } p > 2 \text{ and for } n = 2, 3, \ldots.$$

Assuming (9.1) and combining it with Proposition 9.2(a), we get

$$\gamma_A(l_n^2) \geqslant \frac{k_p(l_n^2)}{k_p(A)} \geqslant C_3 n^{1/2-1/p} \cdot p^{-1} \quad \text{for } p \geqslant 2.$$

Specifying $p = (\log(n+1))^{-1}$, we get $\gamma_A(l_n^2) \geqslant C_4 n^{1/2}/\log(n+1)$.

To establish (9.1) it is enough, in view of Proposition 9.2, to show that there exists a constant C_5 such that the condition (ii) of Proposition 9.1 is violated for $X = l_n^2$ and for every constant $k_p < C_5 n^{1/2-1/p}$. By a recent result of Figiel, Lindenstrauss and Milman **[F-L-M]**, there exists a constant C_6 such that there exists an n-dimensional subspace Y of $l^{p^*}_{C_6 n}$ and an isomorphism U: $Y \longrightarrow l_n^2$ such that $\|U\| = 1$ and $\|U^{-1}\| \leqslant 2$ (note that $p^* = p/(p-1) < 2$). Let $\widetilde{U}$: $l^{p^*}_{C_6 n} \longrightarrow l_n^2$ be any extension of U. Then $(U^{-1}, \widetilde{U})$ is a factorization of l_n^2 through $l^{p^*}_{C_6 n}$. Let I: $l^1_{C_6 n} \longrightarrow l^{p^*}_{C_6 n}$ be the natural embedding. It is easy to see that $\|I\| = 1$ and $\|I^{-1}\| \leqslant (C_6 n)^{1/p} = C_7 n^{1/p}$. Clearly $(I^{-1}U^{-1}, \widetilde{U}I)$ is a factorization of l_n^2. It is well known that for every $L^1(\nu)$ space, $\gamma_{L^1(\nu)}(l_n^2) \geqslant C_8 n^{1/2}$. (This is for instance an easy consequence of Grothendieck's result that every bounded linear operator in l^2 which is L^1-factorable is Hilbert-Schmidt (cf. **[GR1]**, **[L-P1]**).) Thus

$$C_8 n^{1/2} \leqslant \|I^{-1}U^{-1}\| \, \|\widetilde{U} \cdot I\| \leqslant 2n^{1/p}\|\widetilde{U}\|.$$

Hence $\|\widetilde{U}\| \geqslant C_9 n^{1/2-1/p}$. □

Problem 9.1. Does there exist a $C > 0$ such that $\gamma_A(l_n^2) \geqslant C n^{1/2}$?

Observe that the positive answer to Problem 3.1 implies the positive answer to Problem 9.1. On the other hand Theorem 9.1 yields that operators on l^2 which factor through A are "almost" of Hilbert-Schmidt type. We have

COROLLARY 9.1. *If an operator T: $l^2 \longrightarrow l^2$ factors through A, then T is compact and if (s_j) is an enumeration in a nonincreasing sequence of nonzero eigenvalues of $|T|$, each eigenvalue repeated according to its multiplicity, then*

$$\sum \frac{s_j^2}{[\log(j+1)]^c} < +\infty \quad \textit{for every } c > 3.$$

PROOF. Let (U, V) be a factorization of T through A. Since l^2 is reflexive and A has the Dunford-Pettis property (cf. §8), T is compact. Next observe that the polar formula implies that an operator T factors through a Banach space iff its absolute value $|T|$ factors through the same Banach space. Hence without loss of generality one may assume that $T = |T|$ and that the sequence (s_j) is infinite (otherwise there is nothing to prove). Let (e_j) be the orthonormal sequence of eigenvectors of T corresponding to the sequence (s_j) of the eigenvalues. Let us fix a positive integer n and denote by P_n the orthogonal projection from l^2 onto the span $(e_j)_{1 \leqslant j \leqslant n}$. Next define S_n: $P_n(l^2) \longrightarrow P_n(l^2)$ by $S_n(e_j) = s_j^{-1} e_j$ for $j = 1, 2, \ldots, n$ and let U_n be the restriction of U to $P_n(l^2)$. Clearly $S_n P_n V U_n = \mathrm{id}_{P_n(l^2)}$. Hence Theorem 9.1 yields

$$\|S_n\| \, \|P_n\| \, \|V\| \, \|U_n\| \geqslant C \frac{n^{1/2}}{\log(n+1)} \quad \text{for } n = 1, 2, \ldots;$$

since $\|P_n\| = 1$ and $\|U_n\| \leqslant \|U\|$, for $C_1 = C/\|U\|\|V\|$ we have $\|S_n\| \geqslant C_1 n^{1/2}/\log(n+1)$ for $n = 1, 2, \ldots$. Using the relation $s_1 \geqslant s_2 \geqslant \cdots \geqslant s_n$, we easily obtain $\|S_n\| = s_n^{-1}$. Thus $s_n \leqslant C_1^{-1} \log(n+1)/n^{1/2}$. Hence

$$\sum_{n=1}^{\infty} \frac{s_n^2}{[\log(n+1)]^c} \leqslant C_1^{-1} \sum_{n=1}^{\infty} \frac{1}{n[\log(n+1)]^{c-2}} < +\infty$$

for every $c > 3$. □

Another consequence of Theorem 9.1 is

COROLLARY 9.2. *Let $1 < p < \infty$ and let $p^* = p/(p-1)$. Then there exist an absolute constant $C > 0$ and a number $\alpha(p)$ with $0 < \alpha(p) \leqslant \frac{1}{2}$ such that*

$$\gamma_A(l_n^p) = \gamma_{H^\infty}(l_n^p) = \gamma_{L^1/H_0^1}(l_n^{p^*}) \geqslant Cn^{\alpha(p)} \quad \text{for } n = 1, 2, \ldots .$$

PROOF. Combine Theorem 9.1 with a result of Gurariĭ Kadec and Macaev **[G-K-M]** that $d(l_n^2, l_n^p) \leqslant n^{|1/2 - 1/p|}$. □

Next observe that Davis and Johnson **[D-J]** showed that if X is an infinite dimensional Banach space which does not contain l_n^1's uniformly, then there is an α with $0 \leqslant \alpha \leqslant \frac{1}{2}$ such that $\gamma_X(l_n^2) \leqslant Cn^\alpha$ for some absolute constant C. Combining this fact with Theorem 9.1 one can prove

COROLLARY 9.3. *If (X_n) is an increasing sequence of finite dimensional subspaces of a space X with the above property, then*

$$\lim_n \gamma_A(X_n) = \lim_n \gamma_{H^\infty}(X_n) = +\infty.$$

Our next result extends Corollary 9.2 on the case $p = 1$.

THEOREM 9.2. *There is an α with $0 < \alpha \leqslant \frac{1}{2}$ and an absolute constant $C > 0$ such that, for $n = 1, 2, \ldots$,*

$$Cn^\alpha \leqslant \gamma_A(l_n^1) = \gamma_{H^\infty}(l_n^1) = \inf\{d(l_n^\infty, E) | \, E \subset L^1/H_0^1;\ \dim E = n\}.$$

PROOF. It is enough to prove the left-hand inequality. The equalities follow from Proposition 9.1(d) and the duality relations between A, H^∞ and L^1/H_0^1 established in §1. (Note that H^∞ is norm one complemented in A^{**}.) By a result of Maurey and Pisier **[Mau-Pis]**, to prove the inequality it is enough to show that $\lim_n \gamma_A(l_n^1) = \infty$; hence in view of Theorem 2.4 and Proposition 9.1(a), it suffices to show that, for some $p > 2$, $\lim_n k_p(l_n^1) = +\infty$. Assume to the contrary that $\sup_n k_p(l_n^1) = M < \infty$. (Note that the sequence $(k_p(l_n^1))_{n=1,2,\ldots}$ is increasing.) Then, Proposition 9.2 would imply that for every subspace Y of L^{p^*}, every operator $U\colon Y \longrightarrow l_n^1$ admits an extension $\widetilde{U}\colon L^{p^*} \longrightarrow l^1$ with $\|\widetilde{U}\| \leqslant M\|U\|$. Since L^1 contains an increasing sequence of subspaces isometrically isomorphic to l_n^1's whose union is dense in L^1 and since L^1 is complemented in its second dual, a standard compactness argument (for instance involving Banach limits; cf. §5, proof of Proposition 5.2) would yield that every bounded operator from $Y \subset L^{p^*}$ into L^1 extends to a bounded operator from L^{p^*} into L^1. Hence to complete the proof it is enough to establish the following

LEMMA 9.1. *For some $p > 2$ (in fact for every $p > 2$) there exists a bounded linear operator from a subspace of L^{p^*} into L^1 which has no extension to a bounded linear operator from L^{p^*} into L^1.*

PROOF. Let Y be a noncomplemented subspace of L^{p^*} which is isomorphic to l^2 (for instance for every $p > 4$ one can take as Y a closed linear subspace of L^{p^*} spanned by the characters $\{z^n\colon n \in \text{Rud}\}$ where Rud is a Λ_4-set constructed by Rudin, cf. **[Ru2]** and

[R1]; for $4 \geqslant p > 2$ the existence of Y in question follows from the recent result of five authors **[B-D-G-J-N]**). Let E_q $(1 \leqslant q < \infty)$ denote the closed linear space in L^q of the characters $\{z^{2^n}: n = 1, 2, \dots\}$. It is well known (cf. **[Z**, Vol. I, Chapter V, §6**]**) that for every $1 \leqslant q_1 < q_2 < \infty$ the natural embedding $i_{q_2 q_1}: L^{q_2} \longrightarrow L^{q_1}$ maps isomorphically E_{q_2} onto E_{q_1} (in other words all sets of functions E_q coincide with E_2 and the norms $\|\ \|_{q_2}$ and $\|\ \|_{q_1}$ are equivalent on E_2). This implies easily

(9.2) For every $q \geqslant 1$ and $\epsilon > 0$ there exists a $\delta > 0$ such that if e is a measurable subset of ∂D with $m(e) < \delta$ and if $f \in E_2$ with $\|f\|_2 = 1$ then $\int_e |f|^p\, dm < \epsilon$ for $1 \leqslant p \leqslant q$.

Since both Y and E_1 are isomorphic to l^2, there exists an isomorphism $U: Y \xrightarrow[\text{onto}]{} E_1$. We shall show that U is the desired operator. Suppose to the contrary that there exists a bounded linear operator $\widetilde{U}: L^{p^*} \longrightarrow L^1$ which extends U. Then, by a result of Maurey **[Mau2]**, for every fixed r with $p^* > r > 1$, there exists an operator $V: L^{p^*} \longrightarrow L^r$ and a multiplication operator $M_g: L^r \longrightarrow L^1$ defined, for some $g \in L^{r^*}$, by $M_g(f) = gf$ for $f \in L^r$ such that $\widetilde{U} = M_g V$. Next fix q with $r > q > 1$ and using (9.2) pick $\delta > 0$ so small that if e is a measurable subset of ∂D with $m(\partial D \backslash e) < \delta$, then the operator $M_e: L^s \longrightarrow L^s$ of multiplication by the characteristic function χ_e of e restricted to E_s is an isomorphism for $1 \leqslant s \leqslant q^*$. Now we pick $K > 0$ so large that if $e = \{z \in \partial D: |g(z)| \leqslant K\}$ then $m(\partial D \backslash e) < \delta$. By a result of **[K-P]**, the orthogonal projection $Q: L^{q^*} \longrightarrow M_e(E_{q^*})$ is bounded as an operator on L^{q^*}; hence Q being orthogonal is also bounded as an operator from L^q into $M_e(E_q)$. Let $g_e = g \cdot \chi_e$. Clearly the operator M_{g_e} of multiplication by g_e is bounded as an operator from L^r into L^q. Since the restrictions to E_q of M_e and of $i_{q,1}$ are isomorphisms, there is an isomorphism $W: M_e(E_q) \xrightarrow[\text{onto}]{} E_1$ such that $WM_e(f) = i_{q,1}(f)$ for $f \in E_q$. Clearly $WM_{g_e}V(L^{p^*}) \subset E_1$ and $P: U^{-1}WM_{g_e}V$ is a projection from L^{p^*} onto Y, a contradiction. □

Before we pass to a discussion of possible strengthenings of Theorem 9.1 we would like to state a corollary of "global" character which is related to Problem 8.2.

COROLLARY 9.4. *Let X be a subspace of H^∞ satisfying one of the following properties*:

(a) *X is separable and contains an isomorph of c_0.*
(b) *X does not contain l_n^1 uniformly (in particular X is superreflexive)*, $\dim X = \infty$.
(c) *X contains l_n^1 uniformly complemented.*
Then X is uncomplemented in H^∞.

PROOF. (a) Use Sobczyk's theorem **[S]** and a result of **[B-P]** that isomorphs of c_0 are uncomplemented in dual Banach spaces. (b) Use Theorem 9.1 and the result of **[D-J]** mentioned before Corollary 9.3. (c) follows from Theorem 9.2.

REMARK. Using slightly more carefully results of **[B-D-G-J-N]** and **[B-G-N]** one can show that the exponents α in Theorem 9.2 can be made any number $< 1/8$. However, we feel that the "real value" is $\alpha = 1/2$.

Problem 9.2. (a) Does there exist a $C > 0$ such that $\gamma_A(l_n^1) \geqslant Cn^{1/2}$ for every n?

(b) Does every operator from A into l^1 factor through l^2?

(c) Does L^1/H_0^1 have cotype 2, i.e., there is a $C > 0$ such that $\int_0^1 \|\Sigma_{j=1}^n w_j r_j(t)\|_{L^1/H_0^1} dt \geqslant C(\Sigma_{j=1}^n \|w_j\|^2_{L^1/H_0^1})^{1/2}$ for every $w_1, w_2, \ldots, w_n$ in L^1/H_0^1 and for $n = 1, 2, \ldots$? (Here (r_j) denotes the Rademacher functions.)

It follows from a theorem of Maurey (cf. Tomczak Jaegermann **[TJ]**) that the positive answer on (c) implies the positive answer on (b); obviously the positive answer to (b) implies the positive answer on (a).

Recall that a scalar sequence (c_n) is an (A, l^1) *multiplier* if there exists a (unique) bounded linear operator $T_{(c_n)}: A \longrightarrow l^1$ such that $T(z^n) = c_n \delta_n$ for $n = 0, 1, \ldots$ where δ_n denotes the nth unit vector of l^1. It is worthwhile to mention that the answer on Problem 9.2(b) is positive for (A, l^1) multipliers. It is in fact a restatement of an old result of Paley **[Pa2]**. Precisely we have

PROPOSITION 9.3 (PALEY). *If (c_n) is an (A, l^1)-multiplier then $\Sigma |c_n|^2 < +\infty$; hence $T_{(c_n)}$ factors through l^2.*

PROOF. The second part of the proposition is an obvious consequence of the first. We define $U: A \longrightarrow l^2$ by $U(f) = (f(n))$ for $f \in A$ and $V: l^2 \longrightarrow l^1$ by $V((d_n)) = (d_n c_n)$ for $(d_n) \in l^2$. Then clearly $T = VU$.

To prove the first assertion denote by δ_n^* the nth unit vector of c_0 regarded as a functional on l^1 ($n = 0, 1, \ldots$) and note that the adjoint operator $T^*: l^\infty \longrightarrow A^*$ takes δ_n^* into the coset $\{c_n z^{-n} + L^1/H_0^1\}$ for $n = 0, 1, \ldots$ (remember that $A^* = L^1/H_0^1 \dot{\oplus}_1 V_{\text{sing}}$). Fix a positive integer N. By the Kolmogorov theorem (cf. §0.II) there is a constant $C > 0$ such that for an arbitrary sequence (ϵ_j) with $\epsilon_j = \pm 1$ ($j = 1, 2, \ldots, N$) we have

$$\|T^*\| \geqslant \left\| T^*\left(\sum_{j=0}^{N} \epsilon_j \delta_j^*\right)\right\| = \inf_{h \in H_0^1} \int_{\partial D} \left| \sum_{j=0}^{N} c_j \epsilon_j z^{-j} + h(z) \right| m\,(dz)$$

$$\geqslant C \left(\int_{\partial D} \left| \sum_{j=0}^{N} c_j \epsilon_j z^{-j} \right|^{1/2} m(dz) \right)^2.$$

Averaging over all ± 1 sequences, i.e., integrating against the Rademacher functions and applying the Khinchine inequality for $L^{1/2}$ we get

$$C^{-1}\|T^*\| \geqslant \int_0^1 \left(\int_{\partial D} \left| \sum_{j=0}^{N} c_j r_j(t) z^{-j} \right|^{1/2} m(dz) \right)^2 dt$$

$$\geqslant \left(\int_0^1 \int_{\partial D} \left| \sum_{j=0}^{N} c_j r_j(t) z^{-j} \right|^{1/2} m(dz) dt \right)^2$$

$$= \left(\int_{\partial D} \int_0^1 \left| \sum_{j=0}^{N} c_j r_j(t) z^{-j} \right|^{1/2} m(dz)\, dt \right)^2$$

$$\geqslant C_{1/2} \left(\int_{\partial D} \left(\sum_{j=0}^{N} |c_j|^2 \right)^{1/2} m(dz) \right)^2 = C_{1/2} \left(\sum_{j=0}^{N} |c_j|^2 \right)^{1/2}$$

where $C_{1/2}$ is the constant in the Khinchine inequality. Letting $N \longrightarrow \infty$ we get $\Sigma_{j=0}^{\infty}|c_j|^2 < \infty$. □

A positive answer on our last problem will provide a "local" analogue of Corollary 8.

Problem 9.3. Does there exist for $k = 1, 2, \ldots$ and for $C \geqslant 1$ an integer $n = n(k, C)$ such that if X is a subspace of A with $\dim X = n$ which is a range of a projection from A of norm $\leqslant C$, then X contains a k-dimensional subspace, say E with $d(E, l_k^\infty) < 2$?

Notes and remarks to §9. Most of the material of this section was obtained jointly by W. B. Johnson, G. Pisier and the author and is published here for the first time.

Proposition 9.2 is due to Maurey **[Mau1]**. The proof in the text is due to W. B. Johnson and G. Pisier and is published here with their permission.

Proposition 9.3 is due to Paley **[Pa2]**. For other proofs see Helson **[Hel]** and Rudin **[Ru3]**. The proof in the text is similar to Rudin's proof.

ADDED IN PROOF. Every multiplier from L^1/H_0^1 into l^2 is absolutely summing (cf. **[Kw-P2]**). This easily yields Proposition 7.3. It is unknown whether every bounded linear operator from L^1/H_0^1 into l^2 is absolutely summing.

10. Bases and the Approximation Property in Some Spaces of Analytic Functions

In this section we review results on the approximation property and bases in H^p spaces ($1 \leqslant p \leqslant \infty$) and in the disc algebra. Most of the results are stated without proofs. We discuss several open problems.

Recall that a Banach space X has the *approximation property* (shortly A.P.) if for every compact set K in X and $\epsilon > 0$ there exists a finite rank operator $T: X \longrightarrow X$ with $\|T(x) - x\| < \epsilon$ for $x \in K$; moreover, if there exists a $\lambda \geqslant 1$ (which depends on X only but not on K) such that the operator T in question can be chosen with $\|T\| \leqslant \lambda$, then X is said to have the λ-*bounded approximation property* (shortly the λ-B.A.P.). X has the *bounded approximation property* (shortly the B.A.P.) if X has the λ-B.A.P. for some $\lambda \geqslant 1$. A sequence (x_n) of elements of X is called a *basis* for X if for every $x \in X$ there is a unique sequence of scalars (c_n) such that $x = \Sigma c_n x_n$. The basis (x_n) is *unconditional* if the series $\Sigma c_n x_n$ converges unconditionally to X for every $x \in X$.

Each of the above properties is stronger than the preceding one, i.e.,

$$\text{unconditional basis} \Rightarrow \text{basis} \Rightarrow \text{B.A.P.} \Rightarrow \text{A.P.}$$

The table below shows what is known about the approximation property and the existence of bases for "classical" spaces of analytic functions.

	A.P.	1-B.A.P.	basis	unconditional basis
H^∞	?	?	–	–
A	+	+	+	–
H^p $(1 < p < \infty)$	+	+	+	+
H^1	+	+	+	?
L^∞ / H_0^∞	?	?	–	–
$C/A_0 = C(\partial D)/A_0$	+	+	+	?
L^p / H_0^p $(1 < p < \infty)$	+	+	+	+
L^1 / H_0^1	+	+	+	–

We pass to a discussion of the results contained in the table.

The nonexistence of bases in H^∞ and L^∞ / H_0^∞ is trivial because these spaces are unseparable. The nonexistence of unconditional bases in A and L^1 / H_0^1 follows for instance from the results of §4 because those spaces do not have any G-L l.u.st and the existence of an unconditional basis in a Banach space implies the existence of a G-L l.u.st.

All the separable spaces in the table have 1-B.A.P. To this end first recall that a separable Banach space X has the λ-B.A.P. iff there is a sequence $T_n: X \longrightarrow X$ of finite rank operators such that $\|T_n(x) - x\| \longrightarrow 0$ for every $x \in X$ and $\sup_n \|T_n\| \leqslant \lambda$. For $f \in L^1$ we put

$$T_n(f) = \frac{1}{n} \sum_{k=-n+1}^{n-1} (n - |k|)\hat{f}(k)z^k = \frac{1}{n} \sum_{k=0}^{n-1} \left(\sum_{j=-k}^{k} \hat{f}(j)z^j \right) \qquad (n = 1, 2, \dots).$$

It is easy to see that the T_n's take H^p into H^p for $1 \leq p \leq \infty$, A into A, and induce finite rank operators on L^p/H^p_0 ($1 \leq p \leq \infty$) and C/A_0. The T_n's regarded as operators in each of the above spaces are of norm one. By the Fejer theorem (cf. **[H**, p. 23**]**)

$$\|T_n(f) - f\|_p \longrightarrow 0 \quad \text{for } f \in H^p; \qquad \|T_n(f) - f\|_\infty \longrightarrow 0 \quad \text{for } f \in A,$$

$$\|qT_n(f) - q(f)\|_{L^p/H^p_0} \longrightarrow 0 \quad \text{for } f \in L^p; \qquad \|qT_n(f) - q(f)\|_{C/A_0} \longrightarrow 0 \quad \text{for } f \in C(\partial D).$$

Here $q\colon L^p \longrightarrow L^p/H^p_0$ (resp. $q\colon C(\partial D) \longrightarrow C/A_0$) denotes the quotient map. Thus

PROPOSITION 10.1. *Each of the spaces H^p, L^p/H^p_0 $(1 \leq p < \infty)$, A, C/A_0 has* 1-*B.A.P.*

By a result of Boas **[Bo]** (cf. also **[Kw-P]** and §0.II), the spaces H^p and L^p/H^p_0 are isomorphic to L^p for $1 < p < \infty$. By a result of Marcinkiewicz and Paley (cf., e.g., Burkholder and Gundi **[B-G]**), the Haar system is an unconditional basis for L^p $(1 < p < \infty)$. Thus

PROPOSITION 10.2. *If $1 < p < \infty$, then the spaces H^p and L^p/H^p_0 have unconditional bases.*

The situation in the "limit cases" of H^1, A, L^1/H^1_0 and C/A_0 is much more delicate. The existence of bases in these cases is essentially due to Billard **[Bi]** and Bočkarev **[Bt]**. We shall briefly describe their approach.

Let (g_j) be an orthonormal system in the real space $L^2_R[0, 2\pi]$ with $g_0 \equiv 1/\sqrt{2\pi}$. We define a new orthonormal system consisting of 2π-periodic real even functions in $L^2_R(-\pi, \pi)$ by

$$g^e_j(t) = \begin{cases} g_j(2t) & \text{for } t \in [0, \pi], \\ g_j(-2t) & \text{for } t \in [-\pi, 0], \end{cases} \qquad (j = 0, 1, \dots).$$

Next we put

$$g^\#_0 = \frac{1+i}{\sqrt{2\pi}}, \qquad g^\#_j = \frac{1}{\sqrt{2}}\,(g^e_j + i\mathsf{H}(g^e_j)) \qquad (j = 1, 2, \dots)$$

where

$$\mathsf{H}(g^e_j)(t) = \lim_{\epsilon \to +0} -\frac{1}{\pi} \int_\epsilon^\pi \frac{g^e_j(t+s) - g^e_j(t-s)}{2\, tg(s/2)}\, ds$$

is the Hilbert transform of g^e_j.

It is easy to see that $(g^\#_j)$ is an orthonormal sequence in the complex space $L^2(-\pi, \pi)$, in fact in H^2. (We identify H^2 with the subspace of $L^2(-\pi, \pi)$ spanned by the characters $(e^{int})_{n \geq 0}$). Moreover, $(g^\#_j)$ is an orthonormal basis for H^2 if (g_j) is a complete orthonormal system in L^2_R.

Starting with "nice" orthonormal systems in L^2_R one can construct in that way orthonormal systems which form bases for H^p $(\infty > p \geq 1)$ and for A. In particular we have

PROPOSITION 10.3.(BILLARD [Bi]). *Let $(h_n)_{n=0}^{\infty}$ be the Haar system in $L_R^2[0, 2\pi]$, i.e.,*

$$h_0(t) = (2\pi)^{-1/2}$$

$$h_{2^k+r}(t) = \begin{cases} (2\pi)^{-1/2}2^{k/2} & \text{for } 2r\dfrac{2\pi}{2^{k+1}} \leqslant t < (2r+1)\dfrac{2\pi}{2^{k+1}}, \\ -(2\pi)^{-1/2}2^{k/2} & \text{for } (2r+1)\dfrac{2\pi}{2^{k+1}} \leqslant t < 2(r+1)\dfrac{2\pi}{2^{k+1}}, \\ 0 & \text{otherwise,} \end{cases}$$

$$(r = 0, 1, \ldots, 2^k - 1;\ k = 0, 1, \ldots).$$

Then $(h_n^{\#})_{n=0}^{\infty}$ is a basis for H^1.

Let (f_n) be the Franklin system, i.e., the orthonormal system in $L_R^2[0, 2\pi]$ which is obtained by the Gram-Schmidt orthogonalization of the sequence $1, \int_0^t h_0(s)\,ds, \int_0^t h_1(s)\,ds, \ldots$ where $(h_n)_{n=0}^{\infty}$ is the Haar system. Then we have

PROPOSITION 10.4 (BOČKAREV [Bt]). *The system $(f_n^{\#})$ is a basis for A.*

Here we identify A with a subspace of $C[-\pi, \pi]$ spanned by the characters $(e^{int})_{n \geqslant 0}$.

A few words on the proof of Proposition 10.4. Since each f_n is a piecewise linear function, $H(f_n^e)$ is continuous and therefore $f_n^{\#} \in A$ for $n = 0, 1, \ldots$. Next recall that the Franklin system is a basis for $C_R[0, 2\pi]$ (for an elegant proof of this fact cf. [C1]) and therefore

(10.1) the sequence (f_n^e) is a basis for the subspace of $C_R[-\pi, \pi]$ consisting of all the even functions.

The crucial point of the proof is to establish the inequality

$$\left\| H\left(\sum_{n=0}^{N} \langle f, f_n^e \rangle f_n^e \right) \right\|_\infty \leqslant K(\|f\|_\infty + \|H(f)\|_\infty) \tag{10.2}$$

for every even $f \in C_R[-\pi, \pi]$ and for $N = 1, 2, \ldots$

where K is an absolute constant. The proof of (10.2) is difficult. It uses a delicate real variable technique involving an analysis of the behavior of the Hilbert transform of piecewise linear functions with nodes in the "first $2^k + r$" dyadic points of the interval $[-\pi, \pi]$ and some results of Ciesielski [C2].

Next it is convenient to introduce the real space

$$A_R = \{(f, g) \in C_R[-\pi, \pi] \times C_R[-\pi, \pi] : g = H(f)\}.$$

Using (10.1) and (10.2) we easily check that the sequence $(f_n^e, H(f_n^e))$ is a basis for the closed linear subspace of A_R which it spans. This subspace consists of all the pairs $(f, g) \in A_R$ with f even. (The last fact is not obvious but not hard to check.) Thus for each $f \in A$ with $\mathrm{Re}(f)$ even there is a real sequence $(c_j(f))_{j \geqslant 0}$ such that

(10.3) $$f = \sum_{j=0}^{\infty} c_j(f) f_j^{\#}$$

(the series converges uniformly on $[-\pi, \pi]$, i.e., in the norm topology of A). Hence if $\operatorname{Re} f$ is an odd function, then

$$f = \sum_{j=0}^{\infty} -ic_j(if) f_j^{\#}$$

because if $\operatorname{Re}(f)$ is odd then $\operatorname{Im}(f)$ is even.

Finally using the identity

$$f(t) = \tfrac{1}{2}[(f(t) + \overline{f(-t)}) + (f(t) - \overline{f(-t)})]$$

we infer that for every $f \in A$ there is a sequence of complex numbers $(c_j(f))_{j \geqslant 0}$ such that (10.3) holds. The sequence $(c_j(f))_{j \leqslant 0}$ is unique because the sequence $(f_j^{\#})$ is orthonormal.

The proof of Proposition 10.3, being similar to that of Proposition 10.4, is even technically simpler.

There are several consequences of Propositions 10.3 and 10.4. First we have

COROLLARY 10.1. *The spaces C/A_0 and L^1/H_0^1 have bases.*

PROOF. The space C/A_0 is a predual of H^1 (cf. §1). Therefore the existence of a basis in C/A_0 follows from Proposition 10.3 and the result of Johnson, Rosenthal and Zippin **[J-R-Z]** that if X^* has a basis then X does. A strengthening of this general result can be also used to show that L^1/H_0^1 has a basis. There exists however a less formal argument. The coefficient functionals of the Bočkarev basis $(f_j^{\#})_{j \geqslant 0}$ form a basis for the subspace of A^* which they span. Since $(f_j^{\#})_{j \geqslant 0}$ is an orthonormal system, the jth coefficient functionals can be identified with the coset $\{\overline{f}_j^{\#} + H_0^1\}$ for $j = 0, 1, \ldots$ (under the natural identification of L^1/H_0^1 with a subspace of A^*; cf. §1). To complete the proof it is enough to show that the closed linear span E of the cosets $(\{\overline{f}_j^{\#} + H_0^1\})_{j \geqslant 0}$ coincides with L^1/H_0^1. To this end, fix $n \geqslant 0$ and let $e^{int} = \Sigma_{j=0}^{\infty} c_j^n f_j^{\#}(t)$ uniformly on $[-\pi, \pi]$. Then $e^{-int} = \Sigma_{j=0}^{\infty} \overline{c}_j^n \overline{f}_j^{\#}(t)$ uniformly on $[-\pi, \pi]$ and therefore also in L^1-norm. Hence $\{e^{-int} + H_0^1\} = \Sigma_{j=0}^{\infty} \overline{c}_j^n \{\overline{f}_j^{\#} + H_0^1\}$ in the norm topology of L^1/H_0^1. This shows that each coset $\{e^{-int} + H_0^1\}$ belongs to E for $n = 0, 1, \ldots$. Combining the last fact with the Fejer theorem we get the desired conclusion. □

The space $A(D^n)$ of all continuous functions on the n-disc analytic in its interior can be regarded as the n weak tensor product power of the disc algebra. By a result of Gelbaum and Gil de Lamadrid **[G-GL]**, the weak tensor product of Banach spaces with bases has a basis. Hence

COROLLARY 10.2. *The space $A(D^n)$ has a basis $n = 1, 2, \ldots$.*

Very little is known on the existence of bases in the spaces $A(U)$ for other bounded closed domains of holomorphy U in $\mathbf{C}^n$ (cf. §11 for the definition). In particular we have

Problem 10.1. Does the space $A(B_n)$ have a basis?

Here $B_n = \{\mathfrak{z} = (z_j) \in C^n \colon \Sigma |z_j|^2 \leqslant 1\}$. Let us observe that if U is a bounded closed circled domain in $\mathbf{C}^n$ then $A(U)$ has the 1-B.A.P. For, define T_n: $A(U) \longrightarrow A(U)$ to be the nth Fejer sum operator, i.e.,

$$T_n(f) = \sum_{j=0}^{n} \frac{n+1-j}{n+1} f^{[n]} \quad \text{for } f \in A(U)\ (n = 0, 1, \ldots)$$

where $f^{[n]}$ denotes the homogeneous part of degree n of the Taylor expansion of f at zero.

Next we turn to a discussion of the interrogation marks in the table. First we state

Problem 10.2. (a) Does the space C/A_0 have an unconditional basis?

(b) Does H^1 have an unconditional basis?

(c) Is H^1 isomorphic to a subspace of a space with an unconditional basis?

(d) Does H^1 have any local unconditional structure?

Obviously the positive answer to (a) implies the positive answer to (b). Furthermore, it has been observed in **[Kw-P]** and independently by Burkholder (unpublished) that the Billard basis $(h_j^{\#})$ for H^1 is not unconditional. We do not know whether the space H^1 is isomorphic to the space "martingale H^1" – Mart H^1 where the last one is defined to be the space of all the functions f in L^1 whose Fourier-Haar expansion $\Sigma_{j=0}^{\infty} c_j(f) f_j$ converges unconditionally in L^1; we admit

$$\|f\|_{\mathrm{Mart}\,H^1} = \sup_{|\epsilon_j|=1} \left\| \sum_{j=0}^{\infty} \epsilon_j c_j(f) f_j \right\|_{L^1}.$$

Finally in connection with (b) and (d) it is interesting to mention that no example of a subspace of L^1 without any local unconditional structure is known.

Next we pass to the

Problem 10.3. Does H^∞ have the approximation property (resp. the bounded approximation property)?

The same problem is open for L^∞/H_0^∞. The spaces H^∞, L^∞/H_0^∞ and the space of all the bounded linear operators on an infinite dimensional Hilbert space are the most common Banach spaces naturally appearing in analysis for which the approximation property is unknown. All those spaces are nonseparable. However Problem 10.3 can be reduced to some questions on separable function algebras. It is known (cf., e.g., **[J]**) that a Banach space X has A.P. (resp. λ-B.A.P. for some $\lambda \geqslant 1$) iff every separable subspace of X is contained in another separable subspace of X with A.P. (resp. with λ-B.A.P. for the same λ). On the other hand, by a recent result of Marshall **[M]** the Blaschke products are linearly dense in H^∞. Thus the positive answer on Problem 10.3 would follow from the positive answer on the following

Problem 10.4. Does every closed subalgebra of H^∞ generated by a countable set of Blaschke products have A.P. (resp. λ-B.A.P. for some absolute constant λ)?

We do not know the answer to Problem 10.4 even in the case of algebras generated by finitely many Blaschke products. Observe that, by a result of Milne **[Mi]**, there is a separable uniform algebra which fails to have the A.P.

Another way of reducing Problem 10.3 to a question on the disc algebra involves the concept of the uniform approximation property introduced in **[P-R]**. Recall that a Banach space X has the *uniform approximation property* (U.A.P. for short) if there exists a $\lambda < \infty$ and a function $k \longrightarrow N(k)$ $(k = 1, 2, \ldots)$ such that for every subset $F \subset X$ of cardinality k there is an operator $T\colon X \longrightarrow X$ such that $T(f) = f$ for $f \in F$, $\|T\| \leqslant \lambda$, $\dim T(X) \leqslant N(k)$. Obviously the U.A.P. with a constant λ implies λ-B.A.P. Furthermore, a complemented subspace of a Banach space with U.A.P. has the U.A.P. Lindenstrauss and Tzafriri **[L-T1]** proved that a Banach space X has the U.A.P. iff X^{**} does. From the results of §1 it follows that H^∞ is a complemented subspace of A^{**}. Thus

Problem 10.5. Does the disc algebra A have the uniform approximation property?

ADDED IN PROOF. Every finite dimensional in L^1/H^1_0 of norm one is one dimensional; there is no sequence in A and L^1/H^1_0 finite dimensional projections of norm one which tends strongly to the identity operator; in fact the range of every finite dimensional projection in A of norm one is isometrically isomorphic to l^∞_n (cf. **[W3]**.

11. The Polydisc Algebra and the n-Ball Algebra, and Their Duals

Let U be a bounded closed domain of holomorphy in the n-dimensional complex vector space $\mathbf{C}^n$, let ∂U denote the boundary of U. By $A(U)$ we denote the Banach space of all continuous complex valued functions on U which are holomorphic in $U \backslash \partial U$–the interior of U; the norm in $A(U)$ is defined by $\|f\| = \sup_{\mathfrak{z} \in U} |f(\mathfrak{z})| = \sup_{\mathfrak{z} \in \partial U} |f(\mathfrak{z})|$ for $f \in A(U)$. In this section we shall primarily deal with the following domains:

the n-disc, $D^n = \{ \mathfrak{z} = (z_j) \in \mathbf{C}^n \colon \max_{1 \leqslant j \leqslant n} |z_j| \leqslant 1\}$;

the n-ball, $B^n = \{ \mathfrak{z} = (z_j) \in \mathbf{C}^n \colon \Sigma_{j=1}^n |z_j|^2 \leqslant 1\}$.

Clearly $D^1 = B_1 = D$ is the unit disc. The spaces $A(D^n)$ and $A(B_n)$ are called the *n-disc algebra* and the *n-ball algebra* respectively. They are the most natural analogue of the disc algebra among the spaces of analytic functions of several complex variables.

Very little is known about the linear topological classification of the spaces $A(U)$. Clearly if there is a biholomorphic map from the interior of a domain U_1 onto the interior of a domain U_2 which extends to a homeomorphism from U_1 onto U_2, then the spaces $A(U_1)$ and $A(U_2)$ are isomorphic (= linearly homeomorphic). This condition is not necessary as the example shows:

$U_1 = D$–the unit disc in $\mathbf{C}^1$;

$U_2 = D \cup \{z \in \mathbf{C}^1 \colon |z - 3| \leqslant 1\}$–the union of two disjoint discs.

A reasonable conjecture says that the topological dimension of U, i.e., the number of variables, is a linear topological invariant of $A(U)$. More precisely, if U_1 and U_2 are bounded closed domains of holomorphy in $\mathbf{C}^n$ and $\mathbf{C}^m$ respectively and if $n \neq m$, then the Banach spaces $A(U_1)$ and $A(U_2)$ are not isomorphic. On the other hand a beautiful result due to Henkin **[He2]**, **[He3]** shows that the spaces $A(U)$ of the same number of variables may not be isomorphic; in particular, the n-disc algebra is not isomorphic as a Banach space to the n-ball algebra whenever $n \geqslant 2$ (cf. Theorem 11.1 below).

A concrete open question related to the "dimension conjecture" mentioned above is the following

Problem 11.1. *Are the spaces $A(D^n)$ and $A(D^m)$ (resp. $A(B_n)$ and $A(B_m)$) not isomorphic whenever $n \neq m$?*

The next theorem summarizes what is known in the linear topological classification of the Banach spaces $A(D^n)$ and $A(B_m)$.

THEOREM 11.1. (a) (*Henkin* **[He1]**, **[He2]**). *If $n = 1, 2, \ldots$ and $m = 2, 3, \ldots$ then $A(B_n)$ is not isomorphic to $A(D^m)$.*

(b) (*Mitjagin-Pelczynski* **[Mt-P]**). *The disc algebra* $A = A(D^1) = A(B_1)$ *is not isomorphic to* $A(B_n)$ *for* $n \geq 2$.

The isomorphic invariants which enable us to obtain Theorem 11.1 are expressed in terms of the duals of the spaces $A(D^n)$ and $A(B_m)$, and they have been introduced in §1.

We recall those invariants. Let X be a separable Banach space. We say that X has Property I if it is isomorphic to a subspace of a $C(S)$-space whose annihilator in $[C(S)]^*$ is norm separable; X has Property II if X^* is a separable distortion of $L^1(\nu)$, i.e., X^* is isomorphic to the Cartesian product $M \oplus V$ where M is a separable Banach space and V is an $L^1(\nu)$-space.

Clearly the possessing by a Banach space X of Property I (resp. Property II) is an isomorphic invariant of X. Hence our Theorem 11.1 is an immediate consequence of the following.

THEOREM 11.2. *We have*

	Property I	*Property* II
$A = A(D^1) = A(B_1)$	*Yes*	*Yes*
$A(B_n)$ $(n \geq 2)$	*No*	*Yes*
$A(D^n)$ $(n \geq 2)$	*No*	*No*

Most of this section is devoted to the proof of Theorem 11.2. We begin with the "yes" entries of the table. The first row follows easily from the F. and M. Riesz theorem (cf. §1). The last "yes" in the second row is a consequence of Henkin's generalization of the F. and M. Riesz theorem which we present next.

We begin with introducing the concept of an analytic measure (due to Henkin) which plays now an important role in the theory of analytic functions of several complex variables.

Let U be a closed bounded domain of holomorphy in $\mathbf{C}^n$. A sequence (f_j) in $A(U)$ is a *Montel sequence* provided $\sup_j \|f_j\| < \infty$ and $\lim_j f_j(\mathfrak{z}) = 0$ for every $\mathfrak{z}$ in the interior of U.

DEFINITION 11.1. An *analytic measure* on ∂D is a complex Borel measure μ on ∂U such that $\lim_j \int_{\partial U} f_j \, d\mu = 0$ for every Montel sequence (f_j) in $A(U)$. We shall denote by $AM(U)$ the set of all the analytic measures on ∂U.

Clearly we have

PROPOSITION 11.1. (i) *$AM(U)$ is a norm closed subspace of* $[C(\partial U)]^*$.

(ii) $[A(U)]^\perp \subset AM(U)$.

(iii) *If* $\mu \in AM(U)$ *and* $f \in A(U)$ *then* $f\mu \in AM(U)$.

(iv) *If ν is a representing measure for a point* $\mathfrak{z} \in U \backslash \partial U$, *i.e.,* $\int_{\partial U} f \, d\nu = f(\mathfrak{z})$ *for* $f \in A(U)$, *then* $\nu \in AM(U)$. □

The evaluation at a point $\mathfrak{z} \in U \backslash \partial U$ regarded as an element of $[A(U)]^*$ will be denoted by $\varphi_{\mathfrak{z}}^*$.

Now we are ready to state Henkin's result.

THEOREM 11.3. (a) *If* $\nu \in [C(\partial B_n)]^*$ *and* $\nu \ll |\mu|$ *with* $\mu \in AM(\partial B_n)$, *then* $\nu \in AM(\partial B_n)$.

(b) *The dual* $[A(B_n)]^*$ *is isometrically isomorphic to the product* $AM(B_n)/[A(B_n)]^\perp \oplus_1 V$ *where* V *is the* L^1*-space consisting of all measures in* $[C(\partial B_n)]^*$ *which are singular with respect to all positive analytic measures on* ∂B_n.

(c) *The space* $AM(B_n)/[A(B_n)]^\perp$ *is separable and is isometrically isomorphic to the closed linear span of the* $\varphi_{\mathfrak{z}}^*$ *for* $\mathfrak{z} \in B_n \backslash \partial B_n$.

PROOF. The difficult part of the Theorem is (a). Assuming that (a) has been established we deduce (b) and (c) as follows. Let $V = \{\nu \in [C(\partial B_n)]^*: \nu \perp |\mu| \text{ for every } \mu \in AM(\partial B_n)\}$. Then V is a band (= a complex sublattice) in $[C(\partial B_n)]^*$. Hence V is an L^1-space. Furthermore by (a) and the Lebesgue decomposition theorem

$$[C(\partial B_n)]^* = AM(B_n) \oplus_1 V.$$

Factoring through the annihilator $[A(B_n)]^\perp$ and taking into account that the dual of $[A(B_n)]^*$ is naturally isometrically isomorphic (via the restriction map) with the quotient $[C(\partial B_n)]^*/[A(B_n)]^\perp$ and that $[A(B_n)]^\perp \subset AM(B_n)$, we get

$$[A(B_n)]^* = [C(\partial B_n)]^*/[A(b_n)]^\perp = (AM(B_n)/[A(B_n)]^\perp) \oplus_1 V.$$

This completes the proof of (b).

To prove (c) we assume (b) and we shall identify the dual $[A(B_n)]^*$ with the product $AM(B_n)/[A(B_n)]^\perp \oplus_1 V$ and the quotient $AM(B_n)/[A(B_n)]^\perp$ with the subspace $AM(B_n)/A(B_n)^\perp \oplus_1 \{0\}$ of that product. Let E be the norm closed subspace of $[A(B_n)]^*$ spanned by the $\varphi_{\mathfrak{z}}^*$'s for $\mathfrak{z} \in B_n \backslash \partial B_n$. Under the above identification each $\varphi_{\mathfrak{z}}^*$ corresponds to the coset $\{\mu + [A(B_n)]^\perp\}$ where μ is any representing measure for the point $\mathfrak{z} \in B_n \backslash \partial B_n$. Since $\mu \in AM(B_n)$, $\varphi_{\mathfrak{z}}^* \in AM(B_n)/[A(B_n)]^\perp$. Hence $E \subset AM(B_n)/[A(B_n)]^\perp$. Now let $x^* \in [A(B_n)]^* \backslash E$. Then, by the Hahn-Banach theorem, there exists an x^{**} in $[A(B_n)]^{**}$ such that $x^{**}(x^*) = 1$ and $x^{**}(\varphi_{\mathfrak{z}}^*) = 0$ for $\mathfrak{z} \in B_n \backslash \partial B_n$. Thus, by Helly's theorem, for every sequence $(\mathfrak{z}_j)$ in $B_n \backslash \partial B_n$ there is a sequence (f_j) in $A(B_n)$ such that $\varphi_{\mathfrak{z}_j}(f_k) = f_k(\mathfrak{z}_j) = 0$ for $j = 1, 2, \ldots, k$; $x^*(f_k) = 1$; $\|f_k\| \leqslant 2\|x^{**}\|$ $(k = 1, 2, \ldots)$. If we have chosen the sequence $(\mathfrak{z}_j)$ to be dense in $B_n \backslash \partial B_n$, the Montel theorem then yields that (f_j) is a Montel sequence. Thus, if μ is any extension of x^* to a linear functional on $[C(\partial B_n)]^*$, then μ is not an analytic measure because

$$\lim_j \int_{\partial B_n} f_j \, d\mu = \lim_j x^*(f_j) = 1.$$

This shows that $E = AM(B_n)/[A(B_n)]^\perp$.

It remains to prove that E is separable. To this end fix r with $0 < r < 1$ and let Q_r: $A(B_n) \longrightarrow C(rB_n)$ be the restriction map, where $rB_n = \{\mathfrak{z} \in B_n: r^{-1}\mathfrak{z} \in B_n\}$. By the Montel theorem, Q_r is compact. Hence Q_r^*: $[C(rB_n)]^* \longrightarrow [A(B_n)]^*$ is also compact. Thus the set $\{\varphi_{\mathfrak{z}}^*: \mathfrak{z} \in rB_n\}$ is totally bounded being contained in the image under Q_r^* of the unit ball of $[C(rB_n)]^*$. Therefore, the set

$$\{\varphi_{\mathfrak{z}}^*: \mathfrak{z} \in B_n\} = \bigcup_{k=1}^{\infty} \left\{ \varphi_{\mathfrak{z}}^*: \mathfrak{z} \in \frac{k}{k+1} B_n \right\}$$

is separable and so is E. This completes the proof of (c).

PROOF OF (a). It is enough to show

(11.1) if $\mu \in AM(B_n)$, then $\bar{z}_k \mu \in AM(B_n)$ for $k = 1, 2, \ldots, n$.

Indeed (11.1) together with Proposition 11.1(iii) yields $P\mu \in AM(B_n)$ for every polynomial P in the $2n$ variables $z_1, z_2, \ldots, z_n, \bar{z}_1, \bar{z}_2, \ldots, \bar{z}_n$. By the Stone-Weierstrass theorem the polynomials are dense in $L^1(|\mu|)$. Thus $L^1(|\mu|) \subset AM(B_n)$, because $AM(B_n)$ is a closed subspace of $C(\partial B_n)^*$. This proves (a).

Now let m denote the Lebesgue measure on ∂B_n. Before proving (11.1) we need the observation that the assertion of Theorem 11.3(a) holds for $\mu = m$. To this end recall the Cauchy formula for B_n

(11.2) $$f(\mathfrak{z}) = c_n \int_{\partial B_n} \frac{f(\omega)}{[1 - (\mathfrak{z}, \omega)]^n} m(d\omega) \qquad (\mathfrak{z} \in B_n \backslash \partial B_n, f \in A(B_n)),$$

where c_n is a numeric factor independent of f. From (11.2) we derive

$$f(0) = c_n \int_{\partial B_n} f(\omega) m(d\omega),$$

$$\frac{\partial^{k_1 + k_2 + \cdots + k_n}}{\partial z_1^{k_1} \cdots \partial z_n^{k_n}} f(0) = k_1!\, k_2! \cdots k_n!\, c_n \int_{\partial B_n} f(\omega) \prod_{j=1}^{n} \bar{w}_j^{k_j} \cdot m(d\omega)$$

for arbitrary multi-index $(k_1, k_2, \ldots, k_n)$ and for $f \in A(B_n)$. (Here $\omega = (w_1, w_2, \ldots, w_n)$.) Hence the Lebesgue measure m is analytic being a representing measure for the origin. By the Montel theorem, $\lim_j (\partial^{k_1 + \cdots + k_n} / \partial z_1^{k_1} \cdots \partial z_n^{k_n})(f_j(0)) = 0$ for every Montel sequence (f_j) in $A(B_n)$ and every multi-index $(k_1, k_2, \ldots, k_n)$. Therefore, the measures $\Pi_{j=1}^n \bar{z}_j^{k_j} \cdot m$ are analytic and the same application of the Stone-Weierstrass theorem as in the beginning of the proof of (a) yields

(11.3) if ν is a complex Borel measure on ∂B_n which is absolutely continuous with respect to the Lebesgue measure m, then $\nu \in AM(B_n)$.

Next we shall show how to derive (11.1) from (11.2), (11.3) and the following

MAIN LEMMA. *Let*

$$(Tf)(\mathfrak{z}) = c_n \int_{\partial B_n} \frac{\bar{z}_1 - \bar{w}_1}{(1 - (\mathfrak{z}, \omega))^n} f(\omega)\, m(d\omega) \qquad (\mathfrak{z} \in B_n \backslash \partial B_n, f \in A(B_n)).$$

Then Tf extends to a continuous function on $C(B_n)$ for each $f \in A(B_n)$, and moreover the operator $T: A(B_n) \to C(B_n)$ is compact.

PROOF OF (11.1). Without loss of generality one may assume that $k = 1$. Let us set

$$Sf = \bar{z}_1 f - Tf \quad \text{for } f \in A(B_n).$$

Then, by the first part of the Main Lemma, $Sf \in C(B_n)$. Furthermore, by (11.2),

$$(Sf)(\mathfrak{z}) = c_n \int_{\partial B_n} \frac{\overline{w}_1 f(\omega)}{[1 - (\mathfrak{z}, \omega)]^n} m(d\omega) \quad \text{for } \mathfrak{z} \in B_n \backslash \partial B_n.$$

Therefore, $(Sf)(\mathfrak{z})$ depends analytically on $\mathfrak{z}$ in the interior of the ball B_n; hence Sf is holomorphic in $B_n \backslash \partial B_n$, equivalently $Sf \in A(B_n)$. Thus $S\colon A(B_n) \longrightarrow A(B_n)$ is a bounded linear operator. Next observe that for every fixed $\mathfrak{z} \in B_n \backslash \partial B_n$ the measure $\nu_{\mathfrak{z}} \in [C(\partial B_n)]^*$ defined by $\nu_{\mathfrak{z}}(A) = \int_A (\overline{w}_1/[1 - (\mathfrak{z}, \omega)]^n)\, m(d\omega)$ for measurable $A \subset \partial B_n$ is absolutely continuous with respect to m. Hence, by (11.3), $\nu_{\mathfrak{z}} \in AM(B_n)$. Therefore, for every Montel sequence (f_j) in $A(B_n)$ and for every $\mathfrak{z} \in B_n \backslash \partial B_n$

(11.4) $$(Sf_j)(\mathfrak{z}) \longrightarrow 0 \quad \text{as } j \longrightarrow \infty,$$

and

(11.5) $$(Tf_j)(\mathfrak{z}) \longrightarrow 0 \quad \text{as } j \longrightarrow \infty.$$

By the second part of the Main Lemma, (Tf_j) is a sequence of equicontinuous functions (because $\sup_j \|f_j\| < \infty$) which, by (11.5), converges to zero on a dense subset of B_n. Hence

(11.6) $$Tf_j(\mathfrak{z}) \longrightarrow 0 \quad \text{as } j \longrightarrow \infty \text{ uniformly for } \mathfrak{z} \in B_n.$$

Thus, for every $\mu \in [C(\partial B_n)]^*$, $\lim_j \int_{\partial B_n} Tf_j\, d\mu = 0$. Furthermore, for every analytic measure μ,

$$\lim_j \int_{\partial B_n} (Sf_j)\, d\mu = 0$$

because, by (11.4), if (f_j) is a Montel sequence so is (Sf_j). Therefore, for every Montel sequence (f_j) in $A(B_n)$

$$\lim_j \int_{\partial B_n} f_j(\omega)\overline{w}_1\, \mu(d\omega) = \lim_j \int_{\partial B_n} Tf_j\, d\mu + \lim_j \int_{\partial B_n} Sf_j\, d\mu = 0$$

which completes the proof of (11.1).

PROOF OF THE MAIN LEMMA. Let us set $K(\mathfrak{z}, \omega) = (\overline{z}_1 - \overline{w}_1)/[1 - (\mathfrak{z}, \omega)]^n$ for $\mathfrak{z} = (z_1, \ldots, z_n) \in B_n$ and $\omega = (w_1, \ldots, w_n) \in \partial B_n$ and $\mathfrak{z} \neq \omega$; $K(\mathfrak{z}, \omega) = 0$ for $\mathfrak{z} = \omega \in \partial B_n$. First observe that it is enough to show

(11.7) $$\lim_j \alpha_j = 0 \quad \text{where } \alpha_j = \sup_{\mathfrak{z} \in B_n} \int_{A_j(\mathfrak{z})} |K(\mathfrak{z}, \omega)| m(d\omega)$$

$$\text{and } A_j(\mathfrak{z}) = \{\omega \in \partial B_n\colon \operatorname{Re}(\mathfrak{z}, \omega) \geqslant 1 - 1/j\}.$$

Assuming (11.7) we complete the proof as follows. Let $\psi\colon [0, \infty] \longrightarrow [0, 1]$ be a continuous function defined by $\psi(t) = 0$ for $0 \leqslant t < ½$, $\psi(t) = 1$ for $t \geqslant 1$, $\psi(t)$ linear for $½ \leqslant t < 1$. Let

$$K_j(\mathfrak{z}, \omega) = K(\mathfrak{z}, \omega)\psi(j(1 - \operatorname{Re}(\mathfrak{z}, \omega))) \quad \text{for } j = 1, 2, \ldots$$

and let

$$T_jf(\mathfrak{z}) = c_n \int_{\partial B_n} K_j(\mathfrak{z}, \omega)f(\omega)m(d\omega) \qquad (f \in C(B_n),\ \mathfrak{z} \in B_n, j = 1, 2, \ldots).$$

Since the kernels K_j are continuous on $B_n \times \partial B_n$, $T_jf \in C(B_n)$ for $f \in C(B_n)$ and the operators T_j are compact. We have also

$$(11.8) \qquad |T_jf(\mathfrak{z}) - Tf(\mathfrak{z})| \leqslant \alpha_j \sup_{\omega \in \partial B_n} |f(\omega)| \quad \text{for } \mathfrak{z} \in B_n, f \in C(B_n).$$

Hence, by (11.7), the sequence (T_jf) tends uniformly to Tf on B_n. Thus Tf is continuous for every $f \in C(B_n)$ and T is a bounded linear operator on $C(B_n)$. Moreover, $\lim_j \| T_j - T\| = 0$; hence T is compact.

The proof of (11.7) is easy for $n = 1$ because then $|K(\mathfrak{z}, \omega)| = 1$ for $\mathfrak{z} \neq \omega$. If $n \geqslant 2$ we observe first that $|K(\mathfrak{z}, \omega)| \leqslant \|\mathfrak{z} - \omega\|/|1 - (\mathfrak{z}, \omega)|^n$ and $A_j(\mathfrak{z}) \subset A_j(\mathfrak{z}/\|\mathfrak{z}\|)$ for $\mathfrak{z} \neq 0$ and $j = 1, 2, \ldots$. Clearly, by the rotation invariantness of the Lebesgue measure m on ∂B_n, the integral $\int_{A_j(\mathfrak{z}\|\mathfrak{z}\|^{-1})} \|\mathfrak{z} - \omega\|/|1 - (\mathfrak{z}, \omega)|^n\, m(d\omega)$ is a function of $\|\mathfrak{z}\|$ only. Therefore, we may assume without loss of generality that $\mathfrak{z} = ae_1$ where $e_1 = (1, 0, 0, \ldots, 0)$ and $0 < a \leqslant 1$. Hence to prove (11.7) it is enough to show that

$$\lim_{j=\infty} \beta_j = 0 \quad \text{where } \beta_j = \sup_{0 \leqslant a \leqslant 1} \int_{A_j(e_1)} \frac{\|ae_1 - \omega\|}{|1 - a\overline{w}_1|^n} m(d\omega).$$

This will follow if we show

$$(11.9) \qquad \frac{\|ae_1 - \omega\|}{|1 - a\overline{w}_1|^n} \leqslant 2^{n+1/2}|1 - w_1|^{(1-2n)/2} \qquad (0 < a \leqslant 1 \text{ and } \omega \in \partial N_n)$$

and

$$(11.10) \qquad \int_{\partial B_n} |1 - w_1|^{(1-2n)/2}\, m(d\omega) < \infty.$$

To establish (11.9) we need elementary inequalities

$$\left|\frac{a - z}{1 - az}\right| \leqslant 1; \quad \left|\frac{1 - z}{1 - az}\right| \leqslant 2 \qquad (0 < a \leqslant 1, z \in D).$$

The second one yields (for $z = \overline{w}_1$)

$$(11.11) \qquad \frac{1}{|1 - a\overline{w}_1|^{n-1}} \leqslant \frac{2^{n-1}}{|1 - w_1|^{n-1}}.$$

Applying both inequalities we get

$$\left(\frac{\|ae_1 - \omega\|}{|1 - a\overline{w}_1|}\right)^2 = \frac{|a - w_1|^2 + 1 - |w_1|^2}{|1 - aw_1|^2}$$

$$(11.12) \qquad \leqslant \frac{|1 - aw_1|^2 + 1 - |w_1|^2}{|1 - aw_1|^2} \leqslant 1 + \frac{4(1 - |w_1|^2)}{|1 - w_1|^2}$$

$$= \frac{5 - 2\,\mathrm{Re}\, w_1 - 3|w_1|^2}{|1 - w_1|^2} \leqslant \frac{8(1 - \mathrm{Re}\, w_1)}{|1 - w_1|^2} \leqslant 8|1 - w_1|^{-1}.$$

Combining (11.11) with (11.12) we get (11.9).

The proof of (11.10) is routine. After expressing the integral in terms of spherical coordinates (in the unit sphere in R^{2n}) the problem reduces to the finiteness of the integral

$$\int_0^\pi \int_0^\pi \frac{\sin^{2n-2}\alpha \sin^{2n-3}\beta \, d\alpha \, d\beta}{[(1-\cos\alpha)^2 + \sin^2\alpha \cos^2\beta]^{(2n-1)/4}},$$

whose integrand is $\leqslant C/(\alpha^2 + (\beta - \pi/2)^2)^{1/2}$ for some absolute constant C. □

Our next goal is to show that if $n \geqslant 2$ then the spaces $A(D^n)$ and $A(B_n)$ are not isomorphic to complemented subspaces of a $C(S)$-space with a separable annihilator. This will establish the "no" entries in the first row of the table.

We shall need the following concept.

DEFINITION 11.2. A subset F of a closed bounded domain $U \subset \mathbf{C}^n$ is called a *support for the disc algebra* if there exists a map $\varphi_F \colon D \longrightarrow F$ and a function $g_F \in A(U)$ such that

$$\varphi_F(D) = F, \tag{11.13}$$

$$g_F \circ \varphi_F(z) = z \quad \text{for every } z \in D, \tag{11.14}$$

$$|g_F(\omega)| < 1 \quad \text{for every } \omega \in U \backslash (F \cap \partial U), \tag{11.15}$$

$$I_F(f) = f \circ \varphi_F \in A \quad \text{for every } f \in A(U). \tag{11.16}$$

Our next result provides a useful criterion for nonisomorphism of some $A(U)$-spaces to a disc algebra.

THEOREM 11.4. *Let $U \subset \mathbf{C}^n$ be a closed bounded domain of holomorphy. Suppose that*

(*) *there exists in U an uncountable family $(F_\gamma)_{\gamma \in \Gamma}$ of supports of the disc algebra such that*

$$F_\alpha \cap F_\beta \cap \partial U = \varnothing \text{ whenever } \alpha \neq \beta \qquad (\alpha, \beta \in \Gamma).$$

Then the Banach space $A(U)$ is not isomorphic to any subspace of $C(S)$-space with a separable annihilator.

PROOF. Let us set $I_\gamma = I_{F_\gamma}$, $g_\gamma = g_{F_\gamma}$ for $\gamma \in \Gamma$, where the operators I_{F_γ} and the functions g_{F_γ} are the I_F and g_F of Definition 11.2 for $F = F_\gamma$. Let $P_A \colon A \longrightarrow l^2$ be the Paley operator defined by

$$P_A(f) = (\hat{f}(2^k)) \quad \text{for } f \in A,$$

and let $P_\gamma = P_A I_\gamma$ for $\gamma \in \Gamma$. Since the operator P_A is absolutely summing (cf. §3, Example 3.1), so are P_γ and $\pi_1(P_\gamma) \leqslant \pi_1(P_A) = C$ for every $\gamma \in \Gamma$.

Now assume to the contrary that there is a closed linear subspace X of a $C(S)$-space (S compact Hausdorff) such that the annihilator $X^\perp \subset L^1(\lambda)$ for some Borel probability measure λ on S, and there are bounded linear operators $R \colon A(U) \longrightarrow X$ and $Q \colon X \longrightarrow A(U)$ such that $QR = \mathrm{id}_{A(U)}$ is the identity on $A(U)$. Then, by Corollary 2.1, for every $\gamma \in \Gamma$, there exist a nonnegative function h_γ and a compact (even nuclear) operator $V_\gamma: X \longrightarrow l^2$ such that

$$\|P_\gamma Q(x) - V_\gamma(x)\| \leq \int_S |x| h_\gamma \, d\lambda \quad \text{for every } x \in X. \tag{11.17}$$

Next fix ϵ with $0 < \epsilon < (\sqrt{2} - 1)/(\|R\| + 1)$ and, for every $\gamma \in \Gamma$, pick a nonnegative function $b_\gamma \in L^\infty(\lambda)$ so that

$$\int_S |b_j - h_j| \, d\lambda \leq \frac{\epsilon}{2}. \tag{11.18}$$

Let

$$\Gamma_m = \{\gamma \in \Gamma: \|b_\gamma\|_\infty \leq m\} \qquad (m = 1, 2, \ldots).$$

Since $\bigcup_{m=1}^\infty \Gamma_m = \Gamma$ and since Γ is uncountable, at least one of the sets Γ_m, say Γ_{m_0}, is infinite. Now fix a positive integer $M > 4\|Q\| m_0^2$ and distinct indices $\gamma_1, \gamma_2, \ldots, \gamma_M$ in Γ_{m_0}. For simplicity we shall write in the sequel the index j instead of γ_j; for instance P_j instead P_{γ_j}, b_j instead of b_{γ_j}, etc. Let us set

$$y_{j,r} = R((g_j)^{2^{r-1}}) \quad \text{for } j = 1, 2, \ldots, M;\ r = 1, 2, \ldots.$$

Clearly $\|y_{j,r}\| \leq \|R\| \, \|g_j^{2^{r-1}}\|_\infty \leq \|R\|$ because by (11.13), (11.14) and (11.15), $\|g_j^{2^{r-1}}\| = \|g_j\| = 1$. Thus, using the fact that the operators V_j are compact, we extract an infinite increasing subsequence $(r(k))$ of the indices such that

$$\|V_j(x_{j,k})\| < \epsilon \quad \text{for } j = 1, 2, \ldots, M;\ k = 1, 2, \ldots$$

where $x_{j,k} = y_{j,r(2k-1)} - y_{j,r(2k)}$.

Thus, by (11.17) and (11.18),

$$\begin{aligned} \|P_j Q(x_{j,k})\| &\leq \int_S |x_{j,k}| h_j \, d\lambda + \epsilon \leq \int_S |x_{j,k}| b_j \, d\lambda + \|x_{j,k}\| \frac{\epsilon}{2} + \epsilon \\ &\leq \int_S |x_{j,k}| b_j \, d\lambda + (\|R\| + 1)\epsilon \end{aligned} \tag{11.19}$$

because $\|x_{j,k}\| \leq \|y_{j,r(2k-1)}\| + \|y_{j,r(2k)}\| \leq 2\|R\|$ $(j = 1, 2, \ldots, M;\ k = 1, 2, \ldots)$.

On the other hand, using the definition of P_A and remembering that $QR = \mathrm{id}_X$, we get

$$P_j Q(y_{j,r}) = P_j(g_j^{2^{r-1}}) = P_A(z^{2^{r-1}}) = e_r \qquad (j = 1, 2, \ldots, M)$$

where e_r denotes the rth unit vector of l^2 $(r = 1, 2, \ldots)$.

Hence $\|P_j Q(x_{j,k})\| = \sqrt{2}$ because $P_j Q(x_{j,k})$ is a difference of two orthogonal vectors each of norm one. Thus

$$\int_S |x_{j,k}| b_j \, d\lambda \geq \sqrt{2} - \epsilon(\|R\| + 1) \geq 1 \qquad (j = 1, 2, \ldots, M;\ k = 1, 2, \ldots). \tag{11.20}$$

It follows from the assumption (*) that the sets $F_1 \cap \partial U, F_2 \cap \partial U, \ldots, F_M \cap \partial U$ are mutually disjoint. Combining this with (11.13), (11.14) and (11.15), we infer that there exists an index r_0 such that, for $r > r_0$, $\Sigma_{j=1}^M |g_j^{2^{r-1}}(\omega)| \leqslant 2$ for $\omega \in U$. Thus there exists an index k such that for arbitrary complex numbers $c_1, c_2, \ldots, c_M$, we have

$$\left\| \sum_{j=1}^M c_j x_{j,k} \right\| \leqslant \left\| \sum_{j=1}^M c_j y_{j,r(2k-1)} \right\| + \left\| \sum_{j=1}^M c_j y_{j,r(2k)} \right\|$$

$$\leqslant \|Q\| \left(\left\| \sum_{j=1}^M c_j g_j^{2^{r(2k-1)-1}} \right\| + \left\| \sum_{j=1}^M c_j g_j^{2^{r(2k)-1}} \right\| \right)$$

$$\leqslant 4\|Q\| \max_{1 \leqslant j \leqslant M} |c_j|.$$

Hence $\Sigma_{j=1}^M |x_j(s)| \leqslant 4\|Q\|$ for every $s \in U$. Therefore,

$$\sum_{j=1}^M |x_j(s)| b_j(s) \leqslant 4\|Q\| m_0 \quad \text{for } s \in U$$

because $\gamma_j \in \Gamma_{m_0}$ (equivalently $\|b_j\|_\infty \leqslant m_0$). Thus, by (11.14),

$$M \leqslant \sum_{j=1}^M \int_S |x_{j,k}| b_j \, d\lambda = \int_S \sum_{j=1}^M |x_{j,k}(s)| b_j(s) \, d\lambda \leqslant 4\|Q\| m_0$$

which contradicts with the choice of M. □

Now we are ready for

COROLLARY 11.1. *If $n \geqslant 2$ then each of the spaces* (i) $A(B_n)$, (ii) $A(D^n)$ *is not isomorphic to any complemented subspace of a $C(S)$-space with a separable annihilator.*

PROOF. It is enough to show that both domains B_n and D^n $(n \geqslant 2)$ satisfy the assumption of Theorem 11.4.

(i) Let $\Gamma = \{\omega = (w_1, w_2, \ldots, w_n) \in \partial B_n : w_1 = \operatorname{Re} w_1 > 0\}$. For $\omega \in \Gamma$ define $\varphi_\omega : D \longrightarrow B_n$ by $\varphi_\omega(z) = z\omega$ for $z \in D$; next define $g_\omega \in A(B_n)$ by $g_\omega(\mathfrak{z}) = (\mathfrak{z}, \omega) = \Sigma_{k=1}^n z_k \overline{w}_k$ for $\mathfrak{z} = (z_1, z_2, \ldots, z_k) \in B_n$, and put $F_\omega = \varphi_\omega(D)$. Clearly each triple $(F_\omega, \varphi_\omega, g_\omega)$ satisfies the conditions (11.13)–(11.16) of Definition 11.2, and B_n together with the family $(F_\omega)_{\omega \in \Gamma}$ satisfies (*).

(ii) Let $\Gamma = \{\omega = (w_1, w_2, \ldots, w_{n-1}) \in \mathbf{C}^{n-1} : |w_1| = |w_2| = \cdots = |w_{n-1}| = 1\}$. For $\omega \in \Gamma$ define $\varphi_\omega : D \longrightarrow D^n$ by $\varphi_\omega(z) = (w_1, w_2, \ldots, w_{n-1}, z)$ for $z \in D$, and $g_\omega \in A(D^n)$ by $g_\omega(\mathfrak{z}) = z_n \Pi_{k=1}^{n-1} 2^{-1}(z_k + w_k)$ for $\mathfrak{z} = (z_1, z_2, \ldots, z_m) \in D^n$, and put $F_\omega = \varphi_\omega(D)$. Clearly each triple $(F_\omega, \varphi_\omega, g_\omega)$ satisfies the conditions (11.13)–(11.16) of Definition 11.2, and D^n together with the family $(F_\omega)_{\omega \in \Gamma}$ satisfies (*). □

We turn to the last "no" entry in the table of Theorem 11.2. Again we shall prove more than is stated there; namely, we have

THEOREM 11.5. *If $m \geqslant 2$ then the dual $[A(D^m)]^*$ is not isomorphic to any subspace* *of the Cartesian product $M \oplus_1 V$ where M is an arbitrary separable Banach space and V is an $L^1(\nu)$-space.*

This stronger result immediately yields

COROLLARY 11.2. *If $n = 1, 2, \ldots$ and $m = 2, 3, \ldots$ then $A(D^m)$ is not isomorphic to any quotient space of $A(B_n)$.*

Theorem 11.5 itself is an immediate consequence of the following three lemmas stated below.

If E is a Banach space and Γ a set of indices then by $l^1_\Gamma(E)$ we denote the Banach space of all functions $(x_\gamma)_{\gamma\in\Gamma}\colon \Gamma \longrightarrow E$ such that $\|(x_\gamma)_{\gamma\in\Gamma}\| = \sup_{\Gamma_0\subset\Gamma}\Sigma_{\gamma\in\Gamma_0}\|x_\gamma\| < \infty$ where the supremum is extended over all finite subsets Γ_0 of Γ.

LEMMA 11.1. *The space L^1/H^1_0 is not isomorphic to any subspace of an $L^1(\nu)$-space.*

LEMMA 11.2. *If $m \geqslant 2$ then the space $[A(D^m)]^*$ contains a subspace isometrically isomorphic to $l^1_\Gamma(L^1/H^1_0)$ for some uncountable set Γ.*

LEMMA 11.3. *Let E be a Banach space. Assume that for some uncountable set Γ the space $l^1_\Gamma(E)$ embeds isomorphically into a Cartesian product $M \oplus_1 V$ where M is a separable Banach space and V is an $L^1(\nu)$-space. Then E is isomorphic to a subspace of some other $L^1(\mu)$-space.*

PROOF OF LEMMA 11.1. Use Corollary 4.1, and the fact that the dual of L^1/H^1_0 is isometrically isomorphic to H^∞ (cf. §1). □

PROOF OF LEMMA 11.2. Let $\Gamma = \partial D$. To each $w \in \partial D$ we assign the subspace E_w of $[A(D^n)]^*$ consisting of the functionals x^* for which there is a $g_{x^*} \in L^1$ such that

$$(11.21)\qquad x^*(f) = \int_{\partial D} f(z, w, 0, 0, \ldots, 0)\, g_{x^*}(z) m(dz) \quad \text{for } f \in A(D^n).$$

It is clear that the map $x^* \longrightarrow \{H^1_0 + g_{x^*}\}$ is, for every fixed w, an isometric isomorphism from E_w onto L^1/H^1_0. We shall show that the family of these isometric isomorphisms "extends" to an isometric isomorphism from the smallest closed linear subspace of $[A(D^m)]^*$ which contains all the subspaces E_w, for $w \in \Gamma = \partial D$, onto $l^1_\Gamma(L^1/H^1_0)$. To this end it is clearly enough to show that if $\{w_1, w_2, \ldots, w_M\}$ is an arbitrary finite subset of Γ then

$$(11.22)\qquad \left\|\sum_{j=1}^{M} x^*_j\right\| = \sum_{j=1}^{M} \|x^*_j\| \quad \text{for every } x^*_j \in E_{w_j}\ (j = 1, 2, \ldots, M).$$

Let $g_j = g_{x^*_j} \in L^1$ satisfy (11.21) with $x^* = x^*_j$ for $j = 1, 2, \ldots, M$. For $j = 1, 2, \ldots, M$, we define an $h_j \in A(D)$ so that $h_j(w_k) = 0$ for $k \neq j$, $h_j(w_j) = 1$, $|h_j(w)| < 1$ for $w \neq w_j$. The existence of the h_j's with the above properties follows immediately from the Rudin-Carleson theorem (cf. Theorem 2.1). By the definition of the x^*_j's, for $f \in A(D^n)$, for $j = 1, 2, \ldots, M$, and for $r = 1, 2, \ldots$ we have

$$(11.23)\qquad x^*_j(f) = x^*_j(fh^r_j); \quad x^*_j(fh^r_k) = 0 \quad \text{whenever } k \neq j; \quad \|fh^r_j\| \leqslant \|f\|.$$

Here we identify h_j^r with functions in $A(B^m)$ whose value at a point $(z_1, z_2, \ldots, z_n)$ is $[h_j(z_2)]^r$.

Now fix $\epsilon > 0$ and pick $f_j \in A(D^m)$ with $\|f_j\| = 1$ so that $x_j^*(f_j) = \operatorname{Re} x_j^*(f_j) \geqslant \|x_j^*\| - \epsilon/M$ for $j = 1, 2, \ldots, M$. Next pick an integer r so large that

$$\left|\sum_{k=1}^{M} h_k^r(z)\right| \leqslant \sum_{k=1}^{M} |h_k(z)|^r \leqslant 1 + \epsilon \quad \text{for } z \in D. \tag{11.24}$$

Then, by (11.23), putting $x^* = \sum_{j=1}^{M} x_j^*$, we have

$$x^*\left(\sum_{k=1}^{M} h_k^r f_k\right) = \sum_{k=1}^{M} x^*(h_k^r f_k) = \sum_{k=1}^{M}\sum_{j=1}^{M} x_j^*(f_k h_k^r)$$

$$= \sum_{j=1}^{M} x_j^*(f_j h_j^r) = \sum_{j=1}^{M} x_j^*(f_j) \geqslant \sum_{j=1}^{M} \|x_j^*\| - \epsilon,$$

while, by (11.23) and (11.24),

$$\left\|\sum_{k=1}^{M} h_k^r f_k\right\| \leqslant \sup_{z\in D} \sum_{k=1}^{M} |h_k^r(z)|\,\|f_k\| \leqslant 1 + \epsilon.$$

Thus,

$$\sum_{j=1}^{M} \|x_j^*\| \leqslant \|x^*\| \geqslant (1+\epsilon)^{-1}\left|x^*\left(\sum_{k=1}^{M} f_k h_k^r\right)\right| \geqslant \left(\sum_{j=1}^{M}\|x_j^*\| - \epsilon\right)\Big/(1+\epsilon).$$

Letting ϵ tend to 0 we get (11.22). □

PROOF OF LEMMA 11.3. By **[L-P1**, Proposition 7.1**]**, it is enough to show that there is a $C \geqslant 1$ such that for every finite dimensional subspace F of E there is a subspace F_1 in V and an isomorphism Q: $F \longrightarrow F_1$ with $\|Q\|\,\|Q^{-1}\| \leqslant C$. Let q_M and q_V denote the natural projection of $M \oplus_1 V$ onto M and V respectively. Let, for $\alpha \in \Gamma$,

$$E_\alpha = \{(x_\gamma)_{\gamma\in\Gamma} \in l_\Gamma^1(E)\colon x_\gamma = 0 \text{ for } \gamma \neq \alpha\}.$$

Clearly we may identify each E_α with E via the map $(x_\gamma) \longrightarrow x_\alpha$. Let T be an isomorphic embedding of $l_\Gamma^1(E)$ into $M \oplus_1 V$, let T^{-1}: $Tl_\Gamma^1(E) \longrightarrow l_\Gamma^1(E)$ denote the inverse of T and let T_α denote the restriction of T to E_α. Let us fix a finite dimensional subspace F of E and consider the space $B(F, M)$ of all bounded linear operators from F into M. Since M is separable and F finite dimensional, $B(F, M)$ is separable. Thus there is a pair of distinct indices γ_1 and γ_2 such that $\|q_M T_{\gamma_1} - q_M T_{\gamma_2}\| < (2\|T^{-1}\|)^{-1}$ because the set $\{q_M T_\gamma\colon \gamma \in \Gamma\}$ after the identification of each E_γ with E is an uncountable subset of $B(F, M)$.

Next define W: $F \longrightarrow l_\Gamma^1(E)$ by $W(F) = (x_\gamma^f)_{\gamma\in\Gamma}$ where $x_{\gamma_1}^f = -x_{\gamma_2}^f = \tfrac{1}{2}f$ and $x_\gamma^f = 0$ for $\gamma \neq \gamma_1$ and $\gamma \neq \gamma_2$ $(f \in F)$. Clearly W is an isometric isomorphism embedding. Now, let $Q = q_V TW$: $F \longrightarrow V$. Then for every $f \in F$ we have

$$\begin{aligned}\|Qf\| &= \|TWf\| - \|q_M TWf\| \\ &= \|TWf\| - \tfrac{1}{2}\|q_M T_{\gamma_1}(f) - q_M T_{\gamma_1}(f)\| \\ &\geqslant \|T^{-1}\|^{-1}\|Wf\| - \tfrac{1}{2}\|f\|\,\|T^{-1}\|^{-1} \\ &\geqslant \|T^{-1}\|^{-1}\|f\| - \tfrac{1}{2}\|f\|\,\|T^{-1}\|^{-1} \\ &= \tfrac{1}{2}\|T^{-1}\|^{-1}\|f\|,\end{aligned}$$

while $\|Qf\| \leqslant \|q_V\|\|T\|\|W\|\|f\| = \|T\|\|f\|$. Hence $Q\colon F \longrightarrow V$ is an isomorphic embedding with $\|Q\|\|Q^{-1}\| \leqslant 2\|T\|\|T^{-1}\|$ where $Q^{-1}\colon Q(f) \longrightarrow F$ denotes the inverse of Q. This completes the proof. □

Notes and remarks to §11. Theorem 11.3 is due to Henkin **[He2]**. The proof presented in the text is a simplified version of the proof due to Cole and Range **[C-R]**. The proof of the Main Lemma is due to T. Figiel. Similar operators and integrals have been considered in **[Ru4]**.

Theorem 11.3 is a special case of a more general result for strictly pseudoconvex domains. Recall that a bounded closed domain U in $\mathbf{C}^n$ is *strictly pseudoconvex with C^2-smooth boundary* if there is a twice continuously differentiable real function ρ defined on an open neighborhood G of U such that $U = \{\mathfrak{z} \in G\colon \rho(\mathfrak{z}) \leqslant 0\}$, $\partial U = \{\mathfrak{z} \in G\colon \rho(\mathfrak{z}) = 0\}$, grad $\rho \neq 0$ at each point $\mathfrak{z} \in \partial U$, and, for every $\mathfrak{z} \in \partial U$,

$$\sum_{j,k=1}^{n} \frac{\partial^2 \rho(\mathfrak{z})}{\partial z_j \partial \bar{z}_k} w_j \bar{w}_k > 0$$

whenever $0 \neq (w_1, w_2, \ldots, w_n) \in \mathbf{C}^n$ satisfies the condition

$$\sum_{j=1}^{n} \frac{\partial \rho(\mathfrak{z})}{\partial z_j} w_j = 0.$$

Regarding ρ as a function of $2n$ real variables $x_1, y_1, \ldots, x_n, y_n$ where $z_j = x_j + iy_j$ we admit

$$\frac{\partial \rho}{\partial z_j} = \frac{1}{2}\left(\frac{\partial \rho}{\partial x_j} - i\frac{\partial \rho}{\partial y_j}\right); \quad \frac{\partial \rho}{\partial \bar{z}_j} = \frac{1}{2}\left(\frac{\partial \rho}{\partial x_j} + i\frac{\partial \rho}{\partial y_j}\right) \quad (j = 1, 2, \ldots, n).$$

We have

THEOREM 11.3a (CF. HENKIN **[He3]**). *The assertions* (a), (b), (c) *of Theorem* 11.3 *remain valid if B_n is replaced by arbitrary bounded closed strictly pseudoconvex domain with C^2-smooth boundary.*

Theorem 11.3a can be proved along the same line as Theorem 11.3. The main technical difficulty is a construction of an analogue of the Cauchy formula (11.2), i.e., an integral representation

$$f(\mathfrak{z}) = \int_{\partial U} f(\omega) K(\omega, \mathfrak{z}) \sigma(d\omega) \qquad (\mathfrak{z} \in U \backslash \partial U, f \in A(\omega))$$

where σ is the $(2n-1)$-dimensional surface Lebesgue measure and $K: \partial U \times (U \backslash \partial U) \longrightarrow \mathbf{C}$ is a kernel which is analytic in the first variable, and such that for the operator induced by this kernel the analogue of the Main Lemma is true (at this point the "geometry" of the boundary of the strictly pseudoconvex domain is exploited). The desired integral formula has been first constructed by Henkin **[He3]** and by Ramirez de Arellano **[Rm]** for strictly pseudoconvex domains with C^3-smooth boundaries, and next, due to improvement by Øverlid **[Ø]**, for strictly pseudoconvex domains with C^2-smooth boundaries. The reader is referred to the excellent survey **[C-He]** for further information concerning the integral representations and analytic measures.

Also other results of this section can be extended to the case of domains of holomorphy with C^2-smooth boundaries. In particular Corollary 11.1 to Theorem 11.4 admits the following generalization (cf. Mitjagin-Pelczynski **[Mt-P**, Theorem 1**]**).

THEOREM 11.6. *If $n \geqslant 2$ and if $U \subset \mathbf{C}^n$ is a bounded closed domain of holomorphy with the C^2-smooth boundary, then the Banach space $A(U)$ is not complemented in any subspace of a $C(S)$-space with a separable annihilator.*

Before stating a generalization of Theorem 11.5 let us observe that the Shilov boundary of the uniform algebra $A(D^m)$ is the m-torus

$$(\partial D)^m = \{ \mathfrak{z} = (z_j) \in \mathbf{C}^m \colon |z_1| = |z_2| = \cdots = |z_m| = 1\},$$

while the topological boundary of D^m is

$$\partial D^m = \left\{ \mathfrak{z} = (z_j) \in \mathbf{C}^m \colon \max_{1 \leqslant j \leqslant m} |z_j| = 1 \right\}.$$

Hence, if $m \geqslant 2$, then the Shilov boundary of $A(D^m)$ is a proper subset of the topological boundary of D^m.

In general we have

THEOREM 11.7. *Let $U \subset \mathbf{C}^m$ $(m \geqslant 2)$ be a bounded closed domain of holomorphy with C^2-smooth boundary. Assume that the Shilov boundary of the algebra $A(U)$ is a proper subset of the topological boundary of U. Then $[A(U)]^*$ is not a separable distortion of an $L^1(\nu)$-space. Hence $A(U)$ is not isomorphic to $A(U_1)$ for any strictly pseudoconvex domain U_1 in $\mathbf{C}^n$ $(n = 1, 2, \ldots)$ with C^2-smooth boundary.*

Theorem 11.7 is also due to Henkin (unpublished); the second part of Theorem 11.7 has been announced in a slightly weaker form in **[He3]** (cf. Theorem 1.6 of **[He3]**). Theorem 11.5 and Corollary 11.2 are essentially due to Henkin **[He1]**. The proof presented in the text seems to be new.

Let us outline briefly another proof that $A(D)$ is not isomorphic to $A(D^n)$ for $n \geqslant 2$.

Let m_n denote the normalized Haar measure on $[\partial D]^n$ and let $H^p([\partial D]^n)$ denote the closure of $A(D^n)$ in $L^p([\partial D]^n)$, i.e., the completion of $A(D^n)$ in the norm $(\int_{[\partial D]^n} |f|^p \, dm_n)^{1/p}$. Let $i_p^{(n)} \colon A(D^n) \longrightarrow H^p([\partial D]^n)$ be the natural embedding. Clearly $i_p^{(n)}$ is a p-absolutely summing operator with $\pi_p(i_p^{(n)}) = 1$. Next, for $p > 1$, we evaluate the p-integral norm of $i_p^{(n)}$ using the invariance of $i_p^{(n)}$ with respect to action of the group $(\partial D)^n$ on $A(D^n)$ and $H^p([\partial D]^n)$ and the averaging technique exactly in the same way as in the proof of formula (2.17) of §2. We get $i_p(i_p^{(n)}) = \|R^{(n)}\|_p$ where $R^{(n)}$ denotes the orthogonal projection from $L^2([\partial D]^n)$ onto $H^2([\partial D]^n)$ regarded as an operator from $L^p([\partial D]^n)$ onto $H^p([\partial D]^n)$. On the other hand a straightforward argument gives $\|R^{(n)}\|_p = \|R\|_p^n$ where R is the Riesz projection. Thus $i_p(i_p^{(n)}) \geqslant C_1 [p^2/(p-1)]^n$ where C_1 is an absolute constant (cf. §0.II). The above arguement combined with Theorem 2.4 shows that for $n \geqslant 2$, the $(i_p - \pi_p)$-ratio of $A(D^n)$ (cf. Definition 9.2) has a different behavior at ∞ than the $(i_p - \pi_p)$-ratio of $A(D^n)$. In fact we have

$$\lim_{p=\infty} k_p(A(D^n))/k_p(A(D)) = \infty$$

which, in view of Proposition 9.1(a), yields that the Banach space $A(D)$ is not isomorphic to $A(D^n)$ for $n \geqslant 2$. □

REMARK. (1) A similar technique enables us to show $H^\infty(\partial D)$ is not isomorphic to $H^\infty([\partial D]^n)$ for $n \geqslant 2$. (2) However this approach does not seem to work to distinguish between $A(B_n)$ and $A(D)$. The reason is that the norm of the analogue of the Riesz projection–the rotation invariant projection from $L^p(\partial B_n)$ onto $H^p(\partial B_n)$–is of order $Cp^2/(p-1)$. This can be deduced from the proof due to Koranyi and Vagi **[K-V]** of the boundedness of this projection [B. S. Mitjagin, private communication]. (3) We do not know whether for $n \geqslant 2$ and for $1 < p \neq 2 < \infty$ the $(i_p - \pi_p)$-ratio of $A(D^n)$ (resp. $A(B_n)$) is finite. If it is finite, we shall be able at least for the polydisc algebras to use this fact to solve Problem 11.1. Thus

Problem 11.2. Let $1 < p < \infty$. (a) Is every p-absolutely summing operator from $A(D^n)$ (resp. $A(B_n)$) p-integral?

(b) Is it true that for $n = 2, 3, \ldots,$ there exists an absolute constant C_n such that, for every finite rank operator $T: A(D^n) \longrightarrow E$ (E arbitrary Banach space), $i_p(T) \leqslant C_n(p^2/(p-1))^n \pi_p(T)$?

(c) Is it true that there is an absolute constant C such that for every finite rank operator $T: A(B_n) \longrightarrow E$ (E arbitrary Banach space), $i_p(T) \leqslant Cp^2/(p-1)$?

References

[A] E. Amar, *Sur un théorème de Mooney relatif aux fonctions analytiques bornées,* Pacific J. Math. **49** (1973), 311–314. MR **49** #9601.

[A-L] E. Amar and Aline Lederer, *Points exposés de la boule unité de $H^\infty(D)$*, C. R. Acad. Sci. Paris Sér. A-B **272** (1971), A1449–A1452. MR **44** #788.

[B1] E. Bishop, *A general Rudin-Carleson theorem,* Proc. Amer. Math. Soc. **13** (1962), 140–143. MR **24** #A3293.

[B2] ———, *A generalization of the Stone Weierstrass theorem,* Pacific J. Math. **11** (1961), 777–783. MR **24** #A3502.

[B-D-G-J-N] G. Bennett, L. E. Dor, V. Goodman, W. B. Johnson and C. M. Newman, *On uncomplemented subspaces of L^p* $(1 < p < 2)$, Israel J. Math. (to appear).

[B-G] D. L. Burkholder and R. F. Gundy, *Extrapolation and interpolation of quasi linear operators on martingales,* Acta Math. **124** (1970), 249–304.

[B-G-N] G. Bennett, V. Goodman and C. M. Newman, *Norms of random matrices,* Pacific J. Math. **59** (1975), 359–365.

[Bi] P. Billard, *Bases dans H et bases de sous espaces de dimension finie dans A,* Proc. Conf., Oberwolfach (August 14–22, 1971), ISNM Vol. 20, Birkhauser, Basel and Stuttgart, 1972.

[Bo] R. P. Boas, Jr., *Isomorphism between H^p and L^p*, Amer. J. Math. **77** (1955), 655–656. MR **17**, 1080.

[B-P] C. Bessaga and A. Pełczyński, *On bases and unconditional convergence of series in Banach spaces,* Studia Math. **17** (1958), 151–164. MR **22** #5872.

[B-Ph] E. Bishop and R. R. Phelps, *A proof that every Banach space is subreflexive,* Bull. Amer. Math. Soc. **67** (1961), 97–98. MR **23** #A503.

[Bt] S. V. Bočkarev, *Existence of a basis in the space of functions in the disk, and some properties of the Franklin system,* Mat. Sb. (N.S.) **95 (137)** (1974), 3–18 = Math. USSR Sbornik **24** (1974), 1–16.

[C1] Z. Ciesielski, *Properties of the orthonormal Franklin system,* Studia Math. **23** (1963), 141–157. MR **28** #419.

[C2] ———, *Properties of the orthonormal Franklin system.* II, Studia Math. **27** (1966), 289–323. MR **34** #3202.

[C-D] I. Cnop and F. Delbaen, *A Dunford-Pettis theorem for $L^1/H^{\infty\perp}$*, J. Functional Analysis **24** (1967).

[C-H] E. M. Čirka and G. M. Henkin, *Boundary-values properties of holomorphic functions of several complex variables,* Sovremennye Problemy Mat. **4** (1975), 13–142. (Russian)

[Chm] J. Chaumat, *Une géneralisation d'un théoreme de Dunford-Pettis,* Univ. Paris XI, U.E.R. Mathématique no. 85, 1974 (preprint).

[Chv] L. Chevalier, *Algèbres de fonctions à orthogonal purement atomique,* Studia Math. **40** (1971), 81–84. MR **46** #6036.

[C-R] B. Cole and R. M. Range, *A-measures on complex manifolds and some applications,* J. Functional Analysis **11** (1972), 393–400. MR **49** #5398.

[C-W] R. R. Coifman and G. Weiss, *H^p spaces and harmonic analysis,* Bull. Amer. Math. Soc. (to appear).

[De1] F. Delbaen, *Weakly compact operators on the disc algebra,* J. Algebra (to appear).

[De2] ———, *The Dunford-Pettis property for certain uniform algebras,* Pacific J. Math. **65** (1976), 29–33.

[D-J] W. J. Davis and W. B. Johnson, *Compact non-nuclear operators,* Studia Math. **51** (1974), 81–85. MR **50** #5514.

[D-P-R] E. Dubinsky, A. Pełczyński and H. P. Rosenthal, *On Banach spaces X for which $\Pi_2(L_\infty, x) = B(L_\infty, x)$,* Studia Math. **44** (1972), 617–648. MR **51** #1350.

[D-SI] N. Dunford and J. T. Schwartz, *Linear Operators.* Vol. I, Interscience, New York, 1958. MR **22** #8302.

[Du] P. L. Duren, *Theory of H^p spaces,* Academic Press, New York and London, 1970. MR **42** #3552.

[D-U] J. Diestel and J. J. Uhl, Jr., *The Radon-Nikodym theorem for Banach space valued measures,* Rocky Mountain J. Math. **6** (1976), 1–46.

[Etch] A. Etcheberry, *Some uncomplemented uniform algebras,* Proc. Amer. Math. Soc. **43** (1974), 323–325. MR **49** #1123.

[Fi] S. D. Fisher, *Exposed points in spaces of bounded analytic functions,* Duke Math. J. **36** (1969), 479–484. MR **41** #4247.

[F-J-T] T. Figiel, W. B. Johnson and L. Tzafriri, *On Banach lattices and spaces having local unconditional structures with applications to Lorentz function spaces,* J. Approximation Theory **13** (1975), 395–412. MR **51** #3866.

[F-L-M] T. Figiel, J. Lindenstrauss and V. Milman, *The dimension of almost spherical sections of convex bodies* (to appear).

———, *The dimension of almost spherical sections of convex bodies,* Bull. Amer. Math. Soc. **83** (1976), 575–578.

[Four1] J. Fournier, *Fourier coefficients after gaps,* J. Math. Anal. Appl. **42** (1973), 255–270. MR **48** #795.

[Four2] ———, *An interpolation problem for coefficients of H_∞ functions,* Proc. Amer. Math. Soc. **42** (1974), 402–408. MR **48** #8806.

[F-S] C. Fefferman and E. M. Stein, *H^p spaces of several variables,* Acta Math. **129** (1972), 137–193.

[G-GL] B. Gelbaum and J. Gil de Lamadrid, *Bases of tensor products of Banach spaces,* Pacific J. Math. **11** (1961), 1281–1286. MR **26** #5394.

[G-J] D. P. Giesy and R. C. James, *Uniformly non-l^1 and B-convex Banach spaces,* Studia Math. **48** (1973), 61–69. MR **48** #11994.

[G-K-M] V. I. Gurariĭ, M. I. Kadec and V. I. Macaev, *On distances between finite-dimensional analogues of L_p-spaces,* Mat. Sb. **70 (112)** (1966), 481–489; English transl., Amer. Math. Soc. Transl. (2) **76** (1968), 207–216. MR **33** #4649.

[G-L] Y. Gordon and D. R. Lewis, *Absolutely summing operators and local unconditional structures,* Acta Math. **133** (1974), 27–48.

[Gl1] I. Glicksberg, *Measures orthogonal to algebras and sets of antisymmetry,* Trans. Amer. Math. Soc. **105** (1962), 415–435. MR **30** #4164.

[Gl2] ———, *Some uncomplemented function algebras,* Trans. Amer. Math. Soc. **111** (1964), 121–137. MR **28** #4383.

[Gm1] T. W. Gamelin, *Uniform algebras,* Prentice-Hall, Englewood Cliffs, N. J., 1969.

[Gm2] ———, *Restriction of subspaces of $C(X)$,* Trans. Amer. Math. Soc. **112** (1964), 278–286. MR **28** #5331.

[Gr1] A. Grothendieck, *Résumé de la théorie metrique des produits tensoriels topologiques,* Bol. Soc. Mat. São Paulo **8** (1953), 1–79. MR **20** #1194.

[Gr2] ———, *Sur les applications linéaires faiblement compactes d'espaces du type $C(K)$,* Canad. J. Math. **5** (1953), 129–173. MR **15**, 438.

[Gr3] ———, *Produits tensoriels topologiques et espaces nucléaires,* Mem. Amer. Math. Soc. No. 16, 1955. MR **17**, 763.

[H] K. Hoffman, *Banach spaces of analytic functions,* Prentice-Hall, Englewood Cliffs, N. J., 1962. MR **24** #A2844.

[He1] G. M. Henkin, *Nonisomorphism of certain spaces of functions of different numbers of variables,* Funkcional Anal. i Priložen **1** (1967), no. 4, 57–68 = Functional Anal. Appl **1** (1967), no. 4, 306–315. MR **36** #4331.

[He2] ———, *Banach spaces of analytic functions in a sphere and in a bicylinder are not isomorphic,* Funkcional Anal. i Priložen **2** (1968), no. 4, 82–91 = Functional Anal. Appl. **2** (1968), no. 4, 334–341.

[He3] ———, *Integral representations of functions holomorphic in strictly pseudoconvex domains and some applications,* Mat. Sb. **78 (120)** (1969), no. 4, 611–632 = Math. USSR Sbornik **7** (1969), no. 4, 597–616. MR **40** #2902.

[Hel] H. Helson, *Conjugate series and a theorem of Paley,* Pacific J. Math. **8** (1958), 437–446. MR **20** #5397.

[Hv1] V. P. Havin, *Weak completeness of the space L^1/H_0^1,* Vestnik Leningrad Univ. **13** (1973), 77–81. (Russian)

[Hv2] ———, *The spaces H^∞ and L^1/H_0^1,* Zap. Naučn. Sem. Leningrad. Otdel. Mat. Inst. Steklov. (LOMI) **39** (1974), 120–148. (Russian) MR **50** #965.

[J] W. B. Johnson, *A complementably universal conjugate Banach space and its relation to the approximation problem,* Israel J. Math. **13** (1972), 301–310. MR **48** #4700.

[J-R-Z] W. B. Johnson, H. P. Rosenthal and M. Zippin, *On bases, finite dimensional decompositions and weaker structures in Banach spaces,* Israel J. Math. **9** (1971), 488–506. MR **43** #6702.

[Kh1] J.-P. Kahane, *Best approximation in $L^1(T)$,* Bull. Amer. Math. Soc. **80** (1974), 788–804. MR **52** #3845.

[Kh2] ———, *Another theorem on bounded analytic functions,* Proc. Amer. Math. Soc. **18** (1967), 827–831. MR **35** #5912.

[Kis1] S. V. Kisljakov, *Uncomplemented uniform algebras,* Mat. Zametki **18** (1975), 91–96. (Russian)

[Kis2] ———, *On the conditions of Dunford-Pettis, Pełczyński, and Grothendieck,* Dokl. Akad. Nauk SSSR **225** (1975), 1252–1255 = Soviet Math. Dokl. **16** (1975), 1616–1621.

[K-P] M. I. Kadec and A. Pełczyński, *Bases, lacunary sequences and complemented subspaces in the spaces L_p,* Studia Math. **21** (1961/62), 161–176. MR **27** #2851.

[K-V] A. Korányi and S. Vagi, *Singular integrals on homogeneous spaces and some problems of classical analysis,* Ann. Scuola. Norm. Sup. Pisa Sci. Fis. Mat. **25** (1972), 575–648.

[Kw] S. Kwapień, *On a theorem of L. Schwartz and its applications to absolutely summing operators,* Studia Math. **38** (1970), 193–201. MR **43** #3822.

[Kw-P] S. Kwapień and A. Pełczyński, *Some linear topological properties of the Hardy spaces H^p,* Compositio Math. **33** (1976), 261–288.

[L-P1] J. Lindenstrauss and A. Pełczyński, *Absolutely summing operators in L_p-spaces and their applications,* Studia Math. **29** (1968), 275–326. MR **37** #6743.

[L-P2] ———, *Contributions to the theory of the classical Banach spaces,* J. Functional Analysis **8** (1971), 225–249. MR **45** #863.

[L-R] J. Lindenstrauss and H. P. Rosenthal, *The L_p-spaces,* Israel J. Math. **7** (1969), 325–349. MR **42** #5012.

[L-T1] J. Lindenstrauss and L. Tzafriri, *The uniform approximation property in Orlicz spaces,* Israel J. Math. **23** (1976), 142–155.

[L-T2] ———, *Classical Banach spaces,* Lecture Notes in Math. vol. 338, Springer-Verlag, Berlin and New York, 1973.

[M] D. E. Marshall, *Blaschke products generate H^∞,* Bull. Amer. Math. Soc. **82** (1976), 494–496.

[Mau1] B. Maurey, *Espaces de cotype (p, q) et théorémes de relèvement,* C. R. Acad. Sci. Paris Sér. A-B **275** (1972), A785–A788. MR **49** #11235.

[Mau2] ———, *Théorèmes de factorisation pour les opérateurs linéaires à valeurs dans les espaces L^p,* Astérisque No. 11, Soc. Math. de France, Paris, 1974. MR **49** #9670.

[Mau-Pis] B. Maurey and G. Pisier, *Caractérisation d'une classe d'espaces de Banach par des propriétés de séries aléatoires vectorielles,* C. R. Acad. Sci. Paris Sér. A-B **277** (1973), A687–A690. MR **48** #9352.

[Mi] H. Milne, *Banach space properties of uniform algebras,* Bull. London Math. Soc. **4** (1972), 323–326. MR **48** #847.

[Moo] M. C. Mooney, *A theorem on bounded analytic functions,* Pacific J. Math. **43** (1972), 457–463. MR **47** #2374.

[M-P] E. Michael and A. Pełczynski, *A linear extension theorem,* Illinois J. Math. **11** (1967), 563–579. MR **36** #671.

[Mt-P] B. S. Mitjagin and A. Pełczyński, *On the nonexistence of linear isomorphisms between Banach spaces of analytic functions of one and several complex variables,* Studia Math. **56** (1975), 85–96.

[M-S] B. Maurey et L. Schwartz, *Espaces L^p et applications radonifiantes,* Séminaire Maurey-Schwartz, Année 1972–1973, Exposés 1–27, École Polytechnique, Paris, 1973.

———, *Espaces L^p, applications radonifiantes et géométrie des espaces de Banach,* Séminaire Maurey-Schwartz, Année 1973–1974; Exposés 1–26; ibid., Année 1974–1975, Exposés 1–27, École Polytechnique, Paris, 1974, 1975.

[Ø] N. Øverlid, *Integral representation formulas and L^p estimates for the $\bar{\partial}$-equation,* Math. Scand. **29** (1971), 137–160. MR **48** #2425.

[P1] A. Pełczyński, *Projections in certain Banach spaces,* Studia Math. **19** (1960), 209–228. MR **23** #A3441.

[P2] ———, *On simultaneous extension of continuous functions. A generalization of the theorems of Rudin-Carleson and Bishop,* Studia Math. **24** (1964), 285–304. MR **30** #5184a.

[P3] ———, *Supplement to my paper "On simultaneous extension of continuous functions",* Studia Math. **25** (1964), 157–161. MR **30** #5184b.

[P4] ———, *Uncomplemented function algebras with separable annihilator,* Duke Math. J. **33** (1966), 605–612. MR **34** #3373.

[P5] ———, *Some linear topological properties of separable function algebras,* Proc. Amer. Math. Soc. **18** (1967), 652–660. MR **35** #4737.

[P6] ———, *Sur certaines propriétés isomorphiques nouvelles des espaces de Banach de fonctions holomorphes A et H^∞*, C. R. Acad. Sci. Paris Ser. A **279** (1974), 9–12. MR **51** #1359.

[P7] ———, *On Banach space properties of uniform algebras,* Spaces of Analytic Functions (Kristiansand, 1975), Lecture Notes in Math., vol. 512, Springer-Verlag, Berlin and New York, 1976, pp. 109–116.

[P8] ———, *On the impossibility of embedding of the space L in certain Banach spaces,* Colloq. Math. **8** (1961), 199–203. MR **24** #A1008.

[P9] ———, *Banach spaces on which every unconditionally converging operator is weakly compact,* Bull. Acad. Polon. Sci. Sér. Math. Astronom. Phys. **10** (1962), 265–270. MR **26** #6785.

[P10] ———, *p-integral operators commuting with group representations and examples of quasi p-integral operators which are not p-integral,* Studia Math. **33** (1969), 63–70. MR **39** #6125.

[Pal] R. E. A. C. Paley, *On the lacunary coefficients of power series,* Ann. of Math. (2) **34** (1933), 615–616.

[Pa2] R. E. A. C. Paley, *A note on power series,* J. London Math. Soc. 7 (1932), 122–130.

[Per] A. Persson, *On some properties of p-nuclear and p-integral operators,* Studia Math. **33** (1969), 213–222. MR **40** #769.

[Ph1] R. R. Phelps, *Lectures on Choquet's theorem,* Van Nostrand, Princeton, N. J., 1966. MR **33** #1690.

[Ph2] ———, *Extreme points in function algebras,* Duke Math. J. **32** (1965), 267–277. MR **31** #3890.

[Pis] G. Pisier, *Embedding spaces of operators into certain Banach lattices,* Univ. of Illinois at Urbana-Champaign, 1976 (preprint).

[Pi] A. Pietsch, *Absolut p-summierende Abbildungen in normierten Räumen,* Studia Math. **28** (1966/67), 333–353. MR **35** #7162.

[P-P] A. Persson und A. Pietsch, *p-nukleare und p-integrale Abbildungen in Banachräumen,* Studia Math. **33** (1969), 19–62. MR **39** #4645.

[P-R] A. Pełczyński and H. P. Rosenthal, *Localization techniques in L^p-spaces,* Studia Math. **52** (1974/75), 263–289. MR **50** #14174.

[P-S-W] G. Piranian, A. Shields and J. H. Wells, *Bounded analytic functions and absolutely continuous measures,* Proc. Amer. Math. Soc. **18** (1967), 818–826. MR **35** #5911.

[R1] H. P. Rosenthal, *Projections onto translation-invariant subspaces of $L_p(G)$,* Mem. Amer. Math. Soc. No. 63 (1966). MR **35** #2080.

[R2] ———, *On factors of $C[0, 1]$ with nonseparable dual,* Israel J. Math. **13** (1972), 513–522.

[Rm] E. Ramirez de Arellano, *Ein Divisionsproblem und Randintegraldarstellungen in der komplexen Analysis,* Math. Ann. **184** (1969/70), 172–187. MR **42** #4767.

[Ru1] W. Rudin, *Projections on invariant subspaces,* Proc. Amer. Math. Soc. **13** (1962), 429–432. MR **25** #1460.

[Ru2] ———, *Trigonometric series with gaps,* J. Math. Mech. **9** (1960), 203–227. MR **22** #6972.

[Ru3] ———, *Fourier analysis on groups,* Interscience, New York, 1962. MR **27** #2808.

[Ru4] ———, *Spaces of type $H^\infty + C$,* Ann. Inst. Fourier (Grenoble) **25** (1975), 99–125. MR **51** #13692.

[S] A. Sobczyk, *Projection of the space (m) on its subspace (c_0),* Bull. Amer. Math. Soc. **47** (1941), 938–947. MR **3**, 205.

[Sch] H. H. Schaefer, *Banach lattices and positive operators,* Grundlehren Math. Wiss., Band 215, Springer-Verlag, Berlin and New York, 1974.

[St] E. L. Stout, *The theory of uniform algebras,* Bogden Quigley, New York, 1971.

[TJ] N. Tomczak-Jaegermann, *The moduli of smoothness and convexity and Rademacher averages of the trace classes S_p $(1 \leqslant p < \infty)$,* Studia Math. **50** (1974), 163–182. MR **50** #8141.

[Va] N. Th. Varopoulos, *A theorem on operator algebras,* Math. Scand. **37** (1975), 173–182.

[Ve] W. A. Veech, *Short proof of Sobczyk's theorem,* Proc. Amer. Math. Soc. **28** (1971), 627–628. MR **43** #879.

[Vin] S. A. Vinogradov, *Interpolation theorems of Banach-Rudin-Carleson and norms of embedding operators for certain classes of analytic functions,* Zap. Naučn. Sem. Leningrad Otdel. Mat. Inst. Steklov. (LOMI) **19** (1970), 6–54. (Russian) MR **45** #4137.

[W1] P. Wojtaszczyk, *On Banach space properties of uniform algebras with separable annihilator,* Bull. Acad. Polon. Sci. Sér. Math. Astronom. Phys. **25** (1977), No. 1.

[W2] ———, *Weakly compact operators from some uniform algebras,* Studia Math. (to appear).

[Z] A. Zygmund, *Trigonometric series.* I, II, Cambridge Univ. Press, London and New York, 1959. MR **21** #6498.

[De3] F. Delbaen, *The Pelczynski property for some uniform algebras,* Studia Math. (to appear).

[Kis3] S. V. Kisljakov, *Uniform algebras as Banach spaces,* Zap. Naučn. Sem. Leningrad Otdel. Mat. Inst. Steklov (LOMI) **65** (1976), 80–89. (Russian).

[Kis4] ———, *On spaces with a "small" annihilator,* Zap. Naučn. Sem. Leningrad Otdel. Mat. Inst. Steklov (LOMI) **65** (1976), 192–195. (Russian).

[Kw-P2] S. Kwapien and A. Pelczynski, *Remarks on absolutely summing translation invariant operators from the disc algebra and its dual into a Hilbert space* (to appear).

[W3] P. Wojtaszczyk, *On projections in spaces of bounded analytic functions with applications,* Studia Math. (to appear).

BCDEFGHIJ–CM–89876543210